HEALTH AND FOOD SCIENCE

Health and Food Science

M.P. Singh

ANMOL PUBLICATIONS PVT. LTD.
NEW DELHI - 110 002 (INDIA)

ANMOL PUBLICATIONS PVT. LTD.
H.O.: 4374/4B, Ansari Road, Darya Ganj,
New Delhi-110 002 (India)
Ph.: 23278000, 23261597
B.O.: No. 1015, Ist Main Road, BSK IIIrd Stage
IIIrd Phase, IIIrd Block
Bangalore - 560 085 (India)
Visit us at: www.anmolpublications.com

Health and Food Science

First Published, 2007

ISBN 81-261-3078-4

PRINTED IN INDIA

Printed at Mehra Offset Press, Delhi.

Contents

Preface

We are what we think. We think what we eat and drink. This is an undeniable fact that our thoughts are determined by the kind of food we take. Food not only serves as a source of energy, it also plays a significant role in the making of a personality.

For a healthy life, a healthy diet is a must. Our body mechanism is controlled by a number of elements, called vitamins, proteins, enzymes, carbohydrates and other simple and complex chemical phenomena. All these elements find their origin in various food articles, such as, milk, meat, egg, vegetable, cereal, etc. Each has its own characteristics. For a healthy diet, different nutrients have to be taken in a proper proportion, because excess of any of these may result in disturbing the system and affecting the health.

A balanced diet is one, which consists of all the necessary nutrients in a proper proportion required for the development and maintenance of the body. It depends upon age, health and occupation. For instance, a diet rich in carbohydrates, vitamins, mineral salts, fat, and proteins should provide about 3000 calories per day for a normal adult. The proportion among carbohydrates, proteins and fat should be 1 : 1 : 4. This all comes under Food Science, which has been dealt with, at length in this work.

Health and Food Science is an admirable step in this direction. This work throws abundant light on every aspect of food science and food chemistry. It must prove to be a proud possession of everyone, who is health-conscious and takes great care thereof.

Students, scholars, academics, intellectuals and the general readers would find it an interesting work, alike.

Your suggestions and feedback is our strength.

— Editor

1

Food Science

Dietics is a science that deals with the adequacy of diets during normal life cycle and modifications required during diseased conditions. Menu planning is the process of planning and scheduling intake of meals for a general or specific individual requirements. The following terms and concepts are widely used and a clear understanding of these terms makes menu planning efficient.

Initial Conditions

Health is defined by the World Health Organisation of the United Nations as the "State of complete physical, mental and social well being and not merely the absence of disease and infirmity."

The essential requisites (or dimensions) of "health" would include the following:

* Achievement of optimal growth and development, reflecting the full expression of one's genetic potential.
* Maintenance of the structural integrity and functional efficiency of body tissues necessary for an active and productive life.
* Mental well-being.

* Ability to withstand the inevitable process of ageing with minimal disability and functional impairment, and
* Ability to combat disease, such as

a. resisting infections (immunocompetence)
b. preventing the onset (and retarding the progress) of degenerative diseases and cancer and
c. resisting the effect of environmental toxins and pollutants.

Nutrient requirement can be defined as the minimum amount of the absorbed nutrient that is necessary for maintaining the normal physiological functions of the body.

Recommended Dietary Allowances (RDA) is defined as the nutrient present in the diet which satisfies the daily requirement of nearly all individuals in a population. These imply addition of safety factor amount to the estimated requirement to cover the variation among individuals, losses during cooking and the lack of precision inherent in the estimated requirement.

Recommended Dietary Allowances = Requirements + Safety factor.

Recommended Dietary Allowances of an individual depends on many factors like age, sex, physical work and physiological stress.

* *Age:* An adult requires more total calories than a child due to larger size of the body and increase in activity. A growing child requires more calories and protein per kilogram of body weight than an adult.
* *Sex:* Females require less calories than males as BMR is lower and size of the body is smaller.
* *Physical Work:* Sedentary workers require less calories and B-vitamins than hard working persons.
* *Physiological Stress:* During pregnancy and lactation period the requirement of nutrients are increased.

Adequate intake (AI) is a nutrient recommendation based

on observed or experimentally determined approximation of nutrient intake by a group of healthy people when sufficient scientific evidence is not available to calculate an RDA.

Balanced diet is one which contains different types of foods in such quantities and proportions so that the need for calories, proteins, minerals, vitamins and other nutrients is adequately met and a small provision is made for extra nutrients to withstand short duration of leanness.

In addition a balanced diet should provide bio-active phyto-chemicals such as dietary fibre, antioxidants and nutraceuticals which have positive health benefits.

A balanced diet should provide around 60-70% of total calories from carbohydrate, 10-12% from protein and 20-25% of total calories from fat.

Importance of Greenery

Dietary fibre is defined as that portion of food derived from plant cells, which is resistant to hydrolysis/digestion and consists of hemicellulose, cellulose, lignins, oligosaccharides, pectins, gums and waxes.

Fibre is helpful in preventing constipation, coronary heart diseases, diabetes mellitus, obesity and colonic cancer.

Whole grain cereals like ragi, wheat and whole pulses like blackgram are good sources of dietary fibre. Green leafy vegetables, fruits and skins and seeds also contribute to the fibre content of the diet.

Right Proportion

Antioxidants: The normal functioning of cells is dependent on a proper balance of pro-oxidant and antioxidants. The former promote the release of oxygen to provide the energy needed for cell functioning. Antioxidants help to neutralise and counteract the deleterious free radicals but the natural mechanisms may prove to be inadequate. Especially in poor environments, in addition to free radicals, generated as part of the natural metabolic process, the body also has to deal with free radical species derived from such environmental contaminants as smoke, carcinogenic chemicals derived from food additives and industrial effluent and radiation.

Free radicals could promote lipid peroxidation of membrane lipids impairing cell function, they could damage cellular proteins and DNA; they could cause peroxidation of circulating lipoprotein lipids thus converting them into the more deleterious form. Cumulative tissue injury thus caused by free radicals is now known to underlie the pathogenesis of such diverse conditions as cancer, atherosclerosis and heart diseases, radiation damage and accelerated ageing.

Enzyme antioxidants are superoxide dismutase, (Mn-containing superoxide dismutase, Cu-zn containing superoxide dismutase), glutathione peroxidase (selenium dependent), glutathione -S-transferase, glutathione reductase and catalase.

Coenzyme antioxidants are vitamin A, C, and E, and ubiquinol (reduced form of coenzyme Q).

Some nutrients such as carotenoids, ascorbic acid and tocopherols possess antioxidant properties thus playing a dual role. Prominent common foods possessing such antioxidant activity are green leafy vegetables, fruits like amla and guava, palm oil and sunflower oil and spices like turmeric and cloves.

Checking Points

Nutraceuticals: The word combines "nutrition" and "pharmaceuticals" to mean that food extracts can be used as preventive drugs or food supplements.

The major phyto-nutrients identified to have nutraceutical properties include terpenes, phytosterol, phenols and theols.

Terpenes represent the largest class of phyto-nutrients. They are found in green foods, soya products and grains.

Carotenoids and limonoids are the two major terpene sub-classes. Carrots, tomatoes, parsley, orange and spinach are rich sources of carotenoids. They act as vitamin A precursor and have preventive action against many eye diseases. They can also prevent the oxidation of vitamin A and E. Limonoids act as antioxidants protecting lung tissues from free oxygen. Citrus peel can act as nutraceuticals supplying limonoids.

Phytosterols form another major class of phyto-nutrients, which are present in all plants. However, the rich sources having nutraceutical functions include yellow vegetables, seeds of pumpkins and yams. This phyto chemical competes with dietary cholesterol for uptake in the intestine thereby blocking cholesterol absorption into the body. It facilitates the excretion of cholesterol from the body. It can also prevent the development of tumour in breast and prostate glands.

Flavonoides are phyto-chemicals that enhance the effectiveness of vitamin C. They can also act against allergies, ulcers, tumours and platelet aggregation. They are effective in controlling high blood pressure and can reduce the risk of oestrogen induced cancer.

Anthocyanidins or flavonals have significant role in collagen protein synthesis. They are important in sports medicine also. Athletes who exercise excessively produce free radicals that are tackled by anthocyanidins.

Isoflavones can prevent tumour growth. Beans and other legumes are rich sources with nutraceutical properties. Important isoflavones, genestein and daidzin found in soyabeans have nutraceutical properties against tumour growth. People who consume soya food rarely experience breast, uterus and prostate cancer.

Theols are a major sulphur containing class of phyto-nutrients. The food sources having theolic nutraceutical properties include garlic, onions, mustard and other cruciferous vegetables like cabbage and turnip. Important sub-classes under theols are glucosinolates, allelic sulphides, indoles, isoprenoids and tocopherols. Cruciferous vegetables form rich sources of glucosinolates. They activate liver detoxification enzymes and can reduce the incidence of tumor.

Allelic sulphides are potent antimutagenic and anti-carcinogenic agents. They protect against tumours and give cardiovascular protection. They can block the activity of toxins produced by bacteria and viruses, thereby building immunity against diseases.

Isoprenoids are active against free radicals in the body. Thus they reduce oxidative damage of tissues. Tocopherols found in grains and palm oil inhibit breast cancer.

Probiotics and Prebiotics: Probiotics are live microbial food or feed supplements that benefit human and animal health by improving the microbial balance in the intestine.

Prebiotics promote the growth of probiotics in the form of specific substrates. Products containing both probiotics and prebiotics are being commercially produced.

System of Team

The five food group plan permits an individual to plan a menu to achieve nutrient intakes as specified by RDA. The five food groups suggested by ICMR are given in Table below.

The Five Food Groups and their Major Nutrients

	Food Group	*Main Nutrients*
1.	Cereal grains and products: Wheat, Ragi, Bajra, Maize, Jowar, Barley, Rice flakes, wheat flour.	Energy, Protein, Invisible fat, Vitamin-B_1 Rice, Vitamin-B_2 Folic acid, Iron, Fibre.
2.	Pulses and Legumes: Bengalgram, Blackgram, Greengram, Redgram, Lentil (whole as well as dals), Cowpea, Peas, Rajmah, Soyabean, Beans.	Energy, Protein, Invisible fat, Vitamin-B_1 Vitamin-B_2 Folic acid, Calcium, Iron, Fibre.
3.	Milk and Meat Products: Milk, Curd, Skimmed Milk, Cheese, Chicken, Liver, Fish, Egg, Meat.	Protein, Fat, Vitamin-B_2, Calcium.

Contd...

	Food Group	Main Nutrients
4.	Fruits and Vegetables: Fruits: Mango, Guava, Tomato, Papaya, Orange, Sweet lime, Water melon.	Carotenoids, Vitamin-C, Fibre, Invisible fat, Vitamin-B_2, Folic acid, Iron.
	Vegetables: (green leafy) Amaranth, Spinach, Gogu, Drumstick leaves, Coriander leaves, Fenugreek leaves.	Carotenoids Vitamin-B_2, Folic Acid, Calcium, Iron, Fibre.
	Other Vegetables: Carrots, Brinjal, Ladies finger, Beans Capsicum, Onion, Drumstick, Cauliflower.	Carotenoids, Folic acid, Calcium, Fibre
5.	Fats and Sugar: Fats: Butter, Ghee, Hydrogenated fat, Cooking oils like groundnut, Mustard,	Energy, Fat, Essential Fatty acids.
	Coconut Sugar: Jaggery and sugar	Energy

The five food group system can be used by health professionals for the following purposes:

1. *Tool for Nutritional Assessment and Screenings:* A brief dietary history system can disclose inadequacies of nutrient from any of the five groups. The information can be the first clue for the possibility of the subject may be at the risk of developing nutritional deficiency.
2. *Tool for Nutritional Counselling:* The dietary history based on the five food group system allows a health team to counsel or teach a patient about nutrition.
3. *Explaining Therapeutic Diets to the Patient:* Therapeutic diets are scientifically based on nutrient composition and groups which can be used in menu planning.
4. *Food Labelling and Surveillance System:* Food groups can be used for food labelling and for nutrition surveillance system.

The Principles

a. *Meeting Nutritional Requirement: A* good menu is one which will not only provide adequate calories, fat and proteins but also minerals, vitamins essential for the physical well being of each member of a family. In a balanced diet the ratio of energy distribution from carbohydrate, protein and fat would be 7:1:2. The diet should contain 'Basic five food groups'.

b. *Meal Pattern Must Fulfill Family Needs:* A family meal should cater to the needs of the different members. A growing adolescent boy may need rich food to satisfy his appetite, whereas a young child may require soft and bland diet. Pregnant women require more greens in the diet. A heavy worker requires more calories and B vitamins than other members of the family. Meal pattern varies with age, occupation and life-style of the family members.

c. *Meal Planning should Save Time and Energy:* Planning of meals should be done in such a way, that the recipes should be simple and nutritious. By using Basic Five Food Groups, time can be saved.

d. *Economic Consideration:* Any meals that are planned, if do not satisfy the budget of the family, it cannot be put into practice. The cost of meals can be reduced by using seasonal foods, by bulk purchasing and substituting greens for fruits.

e. *Meal Plan should give Maximum Nutrients:* Losses of nutrients during processing, cooking should be minimised. Sprouted grams, malted cereals, fermented foods enhance the nutritive value. Good quality protein should be distributed in all meals. Pressure cooker can be used to conserve the nutrients.

f. *Consideration for Individual Likes and Dislikes:* The meal planned should not only meet RDA but also individual preferences, particularly vegetarian or non-vegetarian preferences. If a person does not like particular greens, it can be tried in a different form or substituted by some other equally nourishing food. Food habits and dietary pattern should also be considered. Religion, traditions and customs of the individual should be considered in planning the menu.

g. *Planned Meals should Provide Variety:* If the meals are monotonous it is not consumed. Variety can be introduced in colour, texture and taste, by using different kinds of foods and cooking methods. Variety also helps in meeting the nutritional requirement.

h. *Meals should Give Satiety:* Each meal should have some amount of fat, protein and fibre to get satiety. Meals should be planned in such a way that interval between the meals is also considered.

i. *Availability of Foods:* Menus should include locally available foods. The wide variation in dietary patterns throughout the world depends largely upon the available food supply.

Points to be Considered: For planning the menu the following points should be considered:

* Variety of foods should be used in the menu. No single food has all the nutrients.
* Caloric allowance can be 50±. For all other nutrients minimum RDA must be met.
* Energy derived from cereals should be not more than 75 per cent.
* Whole grain cereals, parboiled grains or malted grains give higher nutritive value.
* It is better to include two cereals in one meal like rice and wheat or millets and rice.
* Flour should not be sieved for Chapati as it reduces bran content.
* To improve the cereal and pulse protein quality minimum ratio of cereal protein to pulse protein should be 4:1. In terms of the grains it will be 8 parts of cereals and one part of pulses.

* Two to 3 servings of pulses should be taken every day.
* Green leafy vegetables can be taken more than one serving, if fruit is not included in the diet.
* Foods rich in fibre should be included in the diet.
* It is better to serve the fruit raw without much cooking or taking juice out of it. Every diet should contain atleast one medium size fruit.
* There should be a minimum milk of 100 ml/day. One to two glasses of milk or curd should be included in a balanced diet.
* Energy derived from fats or oils is 15-20 per cent of total calories and 5 per cent from sugar and jaggery.
* One egg weighs around 40 g. This can be served along with cereal or pulses to improve the quality of protein. Instead, one serving of poultry/fish can also be included in the diet.
* Inclusion of salads or raita not only helps in meeting the vitamin requirements but the meals would be attractive and have high satiety value due to the fibre content.
* Fried foods cannot be planned if oil allowance is less or in low calorie diets.
* One-third of nutritional requirement—at least calories and protein should be met by lunch or dinner.
* If possible, meals should be planned at a time for several days.
* Usually the number of meals would be four and for very young children and patients, number of meals can be more.
* Ideally each meal should consist of all the five food groups.
* For quick calculations food exchange list can be used.
* Balanced diet should meet antioxidant requirement.

* Processed foods contain a variety of food additives. They may not be nutritionally balanced unless fortified.
* Five servings of fruits and vegetables should be included in a day's diet.

Steps Involved: There are three steps involved in planning a menu

Step I—Recommended Dietary Allowances: To calculate balanced diet, as a first step there is a need to know Recommended Dietary Allowances for different age groups prescribed by Nutrition Expert Committee of ICMR. Table below shows Recommended Dietary Allowances for Indians (1989).

Step II—Food List: Food list can be prepared either by using ICMR Tables or Exchange lists.

Using ICMR Tables: As a second step while planning the daily diet the foods are chosen from all five food groups. To make menu planning more convenient ICMR has suggested the portion size and balanced diets for adults and for different age groups. The balanced diets for adults and different age groups are given as multiples of these portion sizes.

Using Exchange Lists: The diet can also be prescribed in terms of exchange lists. Food exchange lists are groups of measured foods of the same calorific value and similar protein, fat and carbohydrate. In the same group one can be substituted for the other. All foods of exchange lists make a specific contribution to a good diet. None of the exchange groups can itself supply all the nutrients needed for a well balanced diet. Each exchange provides 100 kcals.

There is no fat exchange as the calorie value of the recipes include fat that is used. If additional fat or sugar is used, calorie value can be calculated by multiplying with 9 or 4 per gram respectively.

Recommended dietary allowances for Indians—1989

Group	Particulars	Body wt	Net energy	Protein	Fat	Calcium	Iron	Vitamin A: Retinol	Vitamin A: β-carotene	Thiamine	Riboflavin	Nicotinic acid	Pyridoxin	Ascorbic acid	Folic acid	Vit. B_{12}
		kg	Kcal/d	g/d	g/d	mg/d	mg/d	μg/d	μg/d	mg/d	mg/d	mg/d	mg/d	mg/d	μg/d	μg/d
Man	Sedentary work		2425							1.2	1.4	16				
	Moderate work	60	2875	60	20	400	28	600	2400	1.4	1.6	18	2.0	40	100	1
	Heavy work		3800							1.6	1.9	21				
Women	Sedentary work		1875							0.9	1.1	12				
	Moderate work	50	2225	50	20	400	30	600	2400	1.1	1.3	14	2.0	40	100	1
	Heavy work		2925							1.2	1.5	16				
	Pregnant woman		+300	+15	30	1000	38	600	2400	+0.2	+0.2	+2	2.5	40	400	1
	Lactation															
	0-6 months	50	+550	+25	45	1000	30	950	3800	+0.3	+0.3	+4	2.5	80	150	1.5
	6-12 months		+400	+18						+0.2	+0.2	+3				
Infants	0-6 months	5.4	108/kg	2.05/kg		500				55μg/kg	65μg/kg	710μg/kg	0.1	25	25	0.2
	6-12 months	8.6	98/kg	1.65/kg				350	1400	50μg/kg	60μg/kg	650μg/kg	0.4			
Children	1-3 years	12.2	1240	22			12	400		0.6	0.7	8	0.9		30	
	4-6 years	19.0	1690	30	25	400	18	400	1600	0.9	1.0	11		40	40	0.2-1.0
	7-9 years	26.9	1950	41			26	600	2400	1.0	1.2	13	1.6		60	
Boys	10-12 years	35.4	2190	54	22	600	34	600	2400	1.1	1.3	15	1.6	40	70	0.2-1.0
Girls	10-12 years	31.5	1970	57			19			1.0	1.2	13				
Boys	13-15 years	47.8	2450	70	22	600	41	600	2400	1.2	1.5	16	2.0	40	100	0.2-1.0
Girls	13-15 years	46.7	2060	65			28			1.0	1.2	14				
Boys	16-18 years	57.1	2640	78	22	500	50	600	2400	1.3	1.6	17	2.0	40	100	0.2-1.0
Girls	16-18 years	49.9	2060	63			30			1.0	1.2	14				

Source: Gopalan C., B.V. Rama Sastri and S.C. Balasubramanian 1991, Nutritive Value of Indian Foods, National Institute of Nutrition, ICMR, Hyderabad 500 007, India.

Portion size for Menu Plan

Food groups	*Portion* g	*Energy* kcal	*Protein* g	*Carbohydrate* g	*Fat* g
Cereals and Millets	30	100	3.0	20	0.8
Pulses	30	100	6,0	15	0.7
Egg	50	85	7.0	—	7.0
Meat, Chicken or fish	50	100	9	—	7.0
Milk	100	70	3.0	5	3.0
Roots and Tubers	100	80	1.3	19	—
Green leafy vegetables	100	45	3.6	-	0.4
Other vegetables	100	30	1.7	-	0.2
Fruits	100	40	—	10	—
Sugar	5	20	—	5	—
Fats & Oils	5	45	—	—	5

Balanced Diet for Adults—sedentary/moderate/heavy Activity (number of portions)

Food Groups	*Portion*	*Type of work*					
	g	*Sedentary Man*	*Moderate Woman*	*Heavy Man*	*Woman*	*Man*	*Woman*
Cereals and millets	30	14	10	16	12	23	16
Pulses	30	2	2	3	2.5	3	3
Milk	100 ml	3	3	3	3	3	3
Roots & tubers	100	2	1	2	1	2	2
Green leafy vegetables	100	1	1	1	1	1	1
Other vegetables	100	1	1	1	1	1	1
Fruits	100	1	1	1	1	1	1
Sugar	5	5	4	8	5	11	9
Fats and Oils (visible)	5	4	4	7	6	11	8

For non-vegetarians substitute one pulse portion with one portion of egg/meat/chicken/ fish.

Recipes from exchange lists are selected from all groups and energy and protein value are calculated in accordance with the RDA. If specific number of exchange lists are chosen from each group and energy and protein RDA are met, most of the time all the other nutrients are also met.

Step III: Making the Menu: The foods that are listed in step II are converted into the actual recipes and distributed in different meals like breakfast, lunch, evening tea and dinner. Recipes in exchange lists are distributed in different meals.

Balanced diet for infants, children and adolescents (number of portions)

Food Groups	*Portion* g	*Infants* 6-12 months	*Years* 1-3	4-6	7-9	10-12 Girls	10-12 Boys	13-18 Girls	13-18 Boys
Cereals and millets	30	1.5	4	7	9	9	11	10	14
Pulses	30	0.5	1	1.5	2	2	2	2	2
Milk (ml)	100	5*	5	5	5	5	5	5	5
Roots & tubers	100	0.5	0.5	1	1	1	1	1	2
Green leafy vegetables	100	0.25	0.5	0.5	1	1	1	1	1
Other vegetables	100	0.25	0.5	0.5	1	1	1	1	1
Fruits	100	1	1	1	1	1	1	1	1
Sugar	5	5	5	6	6	6	7	6	7
Fats/Oils (Visible)	5	2	4	5	5	5	5	5	5

* Quantity indicates top milk. For breast-fed infants, 200 ml top milk is required.

One portion of pulse may be exchanged with one portion (50 g) of egg/meat/chicken/fish.

For infants introduce egg/meat/chicken/fish around 9 months. Specific recommendations as compared to a sedentary woman:

Children

1-6 years : ½ to 3/4 the amount of cereals,

		pulses and vegetables and extra cup of milk.
7-12 years	:	Extra cup of milk
Adolescent girls	:	Extra cup of milk
Adolescent boys	:	Diet of sedentary man with extra cup of milk.

100 kcal. exchange list

capacity of one standard katori is 130 ml.

Cereal exchange - 1.5 - 3.5 g protein

Idli (big)	1
(medium)	1½
Dosa (small)	1
(big)	½
Phulka	2
Chapati	1
Puri	1½
Rava Idli	1
Veg. Sandwich	¾
Bread Toast (medium)	1½
Plain rice	¾ K
Upma	1½ K
Veg. Noodles	1 K
Naan	⅔
Cheese Sandwich	⅓
Ragi puttu	¾ K
Ragi adai	¾
Pulao	½ K
Bise bela bath	½ K
Tamarind rice	½ K
Curd rice	½ K

Pulse Exchange 3-5 g protein

Sambar	1½ K
Rasam	2½ K
Thickdal	½ K
Thin dal	1 K
Channa masala	½ K
Dry peas sundal	¾ K
Roasted Bengal gram chutney (without coconut)	½ K
Sprouted moong salad	1 heaped katori
Pesarattu	¾
Baked Masala Vada (with negligible amount of fat)	2 nos
Adai	1
Vada	1
Bajji	2

Milk exchange - 4.5 g protein

Milk	150 ml - 1 tea cup
Curd	150 ml - 1 full katori
Cheese	1 ⅓ cube 30g
Paneer	40 g
Butter milk	1 glass - (350 ml)
Milk kheer	½ K

Meat exchange 5 g of protein

Egg omelette	1
Scrambled egg	One egg
Fish Kolambu	⅔ K

Fish fry	1 small piece
Boiled egg with gravy	⅓ to ½ serving
Meat curry	1 serving
Vegetable A exchange	
Curry without coconut and gravy (with simple seasoning)	1 heaped Katori
Plain tomato soup	2 K (1 soup bowl)
Soup with white sauce	1 K (½ soup bowl)
Mint chutney	7½ tsp.
Mixed vegetable chowder	¾ C
Vegetable B exchange	
Roots and tubers curry	½ K
Cutlet	½ K
Fruit exchange	
Apple	1 medium
Banana	1
Orange	2½
Sapota	2 (small)
Fruit salad (no sugar or dressing)	¾ K
Guava	1 (Big)

The Hierarchy

The food guide pyramid is an educational tool that shows the dietary guidelines in the easily understood graphic format. It was originally prepared by the Human Nutrition Information Service and published in 1992 by the U.S. Department of Agriculture.

It is meant for use by the general healthy population as a guide for the amounts and types of foods to be included in the daily diet. The pyramid was designed to help teach the concepts of variety, moderation and the inclusion of food types in appropriate proportion in the total diet. The food guide pyramid can be modified for different age groups.

The numbers in the figure indicates the exchanges required of the group by an adult man doing sedentary work.

Genesis of Food

Vegetarian Diets: Vegetarian diets consist of foods of plant origin. The two classic variants are lacto-ovo-vegetarian and strict vegetarian or vegan.

There are a number of dietary, religious and moral considerations which may lead individuals to limit or modify the vegetarian diet.

In India most of the population are vegetarians due to economic reasons. Vegetarians should include variety of plant foods like grains, legumes, ruts and seeds to meet the nutritional requirement.

Those whose nutrient needs are especially high due to growth, pregnancy, lactation, or recovery from illness can generally meet their nutrient requirements on well planned vegetarian diet containing dairy products.

Foods like greens and gingelly seeds are good calcium but all may not be available due to the presence of oxalates and phybates. Lacto vegetarians can get plenty of calcium from milk and milk products.

Good plant food sources of iron include dried beans, dark green vegetables, dried fruits and iron fortified cereals. To enhance absorption of iron a good source of ascorbic acid with meals is indicated.

Classifications of vegetarian diet and nutritional adequacy is given in Table below.

Classification of Vegetarianism

Classification	*Foods Included*	*Foods Excluded*	*Limiting Nutrients*
Semi-vegetarian	Fruits, grains, legumes, nuts, seeds, vegetables, milk, milk products, eggs, chicken, fish	All other animal meat	
Pesco-vegetarian	Fruits, grains, legumes, nuts, seeds, vegetables, milk, milk products, eggs and fish	Animal meat	
Lacto-ovo-vegetarian	Fruits, grains, legumes, nuts, seeds, vegetables, milk, milk products	Animal meat and eggs	Iron
Lacto-vegetarian	Fruits, grains, legumes, nuts, seeds, vegetables, milk, milk products	Animal meat and eggs	Iron
Ovo-vegetarian	Fruits, grains, legumes, nuts, milk, and products Seeds, vegetables and eggs	Animal meat,	Vitamin D, calcium, riboflavin, iron
Vegan	Fruits, grains, legumes, nuts, seeds, vegetables	All animal products	High quality protein, sufficient calories, vitamin D, B-12, and riboflavin, iron and calcium

Vitamin B_{12} is found in animal foods. Meeting the requirement of vitamin B_{12} is not a concern for most vegetarians who consume dairy products or eggs. It is a concern for vegans and supplement of vitamin B_{12} to their diets is necessary. For vegans, special care must be taken to ensure adequacy of vitamin D, iron and vitamin B_{12} and if necessary supplementation may be necessary.

Vegetarian diet is a good source of fibre. Vegetarians usually have faster gastrointestinal transit time, bulkier stools and low serum cholesterol levels. In general, coronary heart disease, high blood pressure, type II diabetes, obesity and some forms of cancer tend to develop less often in vegetarians than in non-vegetarians. Lacto vegetarian diet is a low risk diet for colon cancer, oesophageal cancer.

2

Ideal Food

Preventing Adulteration: The practice of adulteration of foodstuffs and marketing of substandard or materials of poor quality is widely prevalent in our country. It has been roughly estimated that over 50 per cent of marketed foodstuffs available today are adulterated in one way or another. The percentage varies from region to region.

It is an unpalatable fact for a nation to admit, but there is little doubt that the Indian consumer, unlike his Western counterpart, cannot take the purity of foodstuffs for granted. Despite the existence of legislation to prevent adulteration of foodstuffs, a very large number offered to the consumer today are injurious to health.

One of the causes of adulteration is that there is a wide gap between production and supply of food articles. The temptation for quick gain and easy profits amongst most merchants, slow action by the authorities, the poor enforcement of Prevention of Food Adulteration Act and leniency shown to traders in regard to the enforcement of the laws on food safety, lengthy procedural wrangles and delays, due to various loopholes in our legal system, contribute to the rampant adulteration of foodstuffs. The gullibility of the predominantly illiterate masses coupled with ignorance, apathy and indifference make for easy adulteration. In a developing country, the cost factor is the most crucial point for a large segment of the population, the cheapest product is preferred to the best. There are a large number of mushroom industries to cater to this section.

Almost all foods, milk, cereals, dais, spices, ghee, oils and beverages are adulterated. The adulterant used generally mixes well with the major food article in colour, shape, size and appearance. The more highly priced foods and those foods which are in great demand are the ones more often widely adulterated. By adulterating the foodstuffs, the merchant is benefited in many ways. First more weight is added to the commodity.

Secondly, addition of colour improves the appearance of the product and hence substandard product can fetch higher price. Thirdly, new product when combined with the old one, lowers the quality and at the same time substandard commodity is sold at a higher rate. The ingenuity, imagination and initiative shown in adulteration make it difficult for an average buyer to detect it at first sight. Sometimes, it can be detected only in well-equipped laboratories. Besides the relatively harmless forms of adulteration like addition of water to milk, moisture to butter, mixing of edible oils with cheaper edible oils, many types of adulteration are positively injurious to health.

Rice and pulses are often polished with talc, a silicate containing asbestos fibre, which is not removed even after washing, cleaning and cooking. The Japanese have traced cases of stomach cancer to this practice. The other common adulterants used in cereals such as rice and wheat are clay particles, grit, soap stone and weed seeds. Most yellow pulses are adulterated either with kesari dal or by polishing with metanil yellow, a cheap soluble coal tar dye that makes the pulses deceptively brighter and cleaner looking. Metanil yellow is a highly toxic substance with accumulative tendencies, which over the years build up enough toxicity in the body to cause testicular degeneration in the male. When khesari dal is consumed continuously for a few months, it results in the development of lathyrism, a permanent paralysis of the lower limbs.

Spices, particularly because of their high cost, are the most adulterated item in food commodities. Turmeric is coated with lead chromate or metanil yellow to give a lustrous yellow colour. Black pepper is sometimes mixed with papaya berries. Foodgrains and pulses are often indiscriminately sprayed with insecticides to prevent infestation or to destroy pests; chillies are soaked in soluble coal tar dyes to give them a deeper hue, while sawdust, blackgram husk, and used tea leaves are utilised for adulterating tea dust/leaves. These are a few of the many common practices that are used today. Many cheap sweets, sherbets in variety of colours and ice fruits are adulterated with many non-permitted coal tar dyes.

These undesirable colours can impair the liver and can be the cause of cancers. Similarly, lead chromate when added to turmeric powder can cause lead poisoning resulting in stiffness of limbs and paralysis. Mineral oils and other non-edible oils when mixed with edible oils can induce nausea, purging, impaired liver functions and cancer. Continuous use of mustard oil with more than 11 per cent argemone seed oil causes the disease 'epidemic dropsy' in people. Further, many mould-infested grains and oil seeds are further processed into flours and oils. Here the main danger is due to the mould growth which produces many hidden toxic substances called mycotoxins. These can cause various diseases and they act as nervous and neurotoxins impairing the general health. Some also have been found to be carcinogenic.

The menace of food adulteration exists all over the world. In developed countries, due to the implementation of food laws for the last few decades, there has been a significant improvement in the quality of food. But due to tremendous improvement in food industry and technology, various chemical substances like artificial colours, preservatives, antioxidants, emulsifiers, flavour and taste improvers are increasingly being used.

Some of these pose a great threat to the health of the people. As against this, in developing countries adulteration is found to be rampant even in primary articles of food like milk, ghee, edible oils and foodgrains. Shortage of food and spiralling prices have aggravated the problem. Thus all countries in the world have to grapple in their own way with problems raised by food adulteration.

Legal Dimensions

In India, as early as in 1860, adulteration of food was dealt with in certain sections of the Indian Penal Code (Sec. 272,273). Later it was included under the provisions of the Municipal Act. But as a comprehensive act, it was first promulgated in Bombay State in 1899. Subsequently similar acts were enacted by other states also. In the erstwhile Travancore State an Act—The Travancore Prevention of Adulteration Act—was enacted in 1931. Besides food adulteration, adulteration of other articles was also made an offence under this act But then acts were not comprehensive in many ways.

The punishment prescribed was not deterrent enough to prevent adulteration of food efficiently. There was no uniformity of standards prescribed for various food articles. The necessity of a comprehensive act applicable to all the states was felt. In 1938 the Central Advisory Board of Health was established. This Board in its report stressed the necessity for efficiently tackling the problem by all states. When the Constitution of India was framed the topic of prevention of adulteration of food was brought under the concurrent list and the P.F.A. Act was formulated in 1954. It was enacted on June 1,1955.

After about ten years of implementation of the Act, the position was reviewed and some amendments were made in the Act in 1964 to plug loopholes and to give more protection to the honest traders. Again in 1976 some other important amendments were made to the present act to make it more effective and purposeful. These amendments came into force from April 1,1976.

Object of the Act: What the consumer wants is genuine, wholesome food, of nutritive value, free from any health hazard. He also expects that labelling must be truthful and there shall not be fraudulent claims of any kind by any statement, device or design. The interest of the trader is not only to sell the food articles to the consumers but also to make the maximum profit out of it.

This conflicting interest of the consumer on the one hand and the traders on the other, lead to unhealthy practices in trade. It is to prevent such practices that the Act has been made. Though the consumer's interest is the foremost in the P.FA. Act, that of the traders have also been accommodated reasonably as far as scientific grounds and consumers' interest permit.

Important Provisions of the Act: Food has been defined under the Act as articles other than drugs and water, which are ordinarily used as food on drink and all articles which enter into and are used in the preparation of human food. Another definition of 'primary food' has also been introduced by the amendment act of 1976. Primary foods are products of agriculture or horticulture in its natural form.

The term adulterated has been defined under Section 2 of the Act: An article of food will be deemed to be adulterated if it is not of the nature, substance or quality demanded by the purchaser and is to his prejudice or not of the nature, substance or quality which it purports to be; if an article is partly or wholly substituted by the cheaper substance; if any constituent of the article is extracted from it, if it is prepared or exposed for sale under insanitary and unhygienic surrounding; if it contains prohibited colouring matter or preservative or permitted colouring matters and preservatives in excess of the permissible limits; if it does not conform to the standards under all these conditions it will be deemed to be adulterated.

There is a provision that a primary food will not be deemed to be adulterated if it does not conform to standards due to natural causes beyond the control of human agency. In short, anything which lowers the quality of a food article has been brought wider the definition of adulteration.

Fake Variety

Sale of food articles without proper labelling, imitation of other food articles and coating food articles to give improved appearance are brought wider the definition 'misbranding'.

Prohibitory Section: Section 7 of the P.F.A. Act prohibits the sale of (a) any adulterated food, (b) any misbranded food, (c) any article of food for the sale of which a licence is prescribed except in accordance with the conditions of the licence, (d) any article of food and sale of which has been prohibited by the Food (Health) Authority in the interest of public health, (e) any article of food in contravention of the provision of the Act and Rules, and (f) any adulterant.

Crime and Punishment: Infringement of the provision of Section 7 is punishable under Section 16. Various offences and punishments for each have been detailed under these sections. This section has been amended by the amendment act of 1976. Non-graded punishment, depending on the nature of the offence has been laid down. For the sale of adulterated food the minimum punishment is six month's imprisonment and a fine of rupees one thousand.

If the adulterant is of such a nature as to cause death or grievous hurt when consumed by a person, the punishment is imprisonment for a term of three years which may extend to life imprisonment and a fine which shall not be less than rupees live thousand.

Punishment has been prescribed for sale of misbranded food and the sale of adulterant. Preventing a Food Inspector from exercising his powers and duties under the Act, using a report of the Public Analyst of Director of Central Food Laboratory for advertising purposes, not issuing warranty, issuing false warranty, tampering of seized food articles are all punishable under the law.

Protection of the Traders: Section 7 of the Act prohibits the sale of adulterated food or any consumable article, by any person either by himself or by any person on his behalf. Prima facie it may seem that an innocent person who buys a food article from a wholesaler or manufacturer and sells it without knowing that it is adulterated is punishable under the Act. Sections 14, 14 A, 19(2) and 20A are meant to protect honest vendors of food articles.

Section 14 lays down that no manufacturer, distributor or dealer of any article of food shall sell such article to any vendor unless he also gives a warranty in writing about the nature of substance and quality of such food to the vendor.

Insistence of warranty at the time of sale by the manufacturer, distributor or dealer would make it easy to pin down the responsibility with the offender. The amendment of 1976 makes this easier by incorporating a provision that a cash bill can also be considered as a warranty.

According to Section 14A, the vendor has to disclose to the Food Inspector the name, address and other particulars of the person from whom he purchased the food article. This helps the Food Inspector to proceed against the warrantor also. According to Section 19(2) of die Act, a vendor shall not be deemed to have committed an offence if he proves that he purchased the food article with a proper warranty, that he stored the article properly and that he sold it in the same condition in which it was purchased.

Section 19(3) confers a right on the warrantor to appear in court and give evidence. The law thus protects the warrantor also against spurious warranties.

When warranty is proved, the vendor can be exonerated and the warrantor can be proceeded against. As per Section 20A during the trial, when warranty is proved the court can proceed against the warrantor as though a prosecution has been launched against him.

Quality Maintenance

There are two standards set up by the government to help the manufacturers, the traders and the consumers to ensure quality. They are 'Agmark' standard and 'Indian Standards Institution mark.

Agmark Standard: The Directorate of Marketing and Inspection of the Government of India have set up this standard. This grades goods into (1) Special (Grade I), (2) Good (Grade II), (3) Fair (Grade III) and Ordinary (Grade IV).

These standards also specify the type of packaging to be used for different rent products. An Agmark certificate is issued to traders and the mark on commodities provides assurance of the quality to the consumer.

Indian Standards Institution (ISI): The ISI mark on food is a guarantee of good quality. This institution prescribes standards for various food items and issues certificates, to traders.

Common Food Adulterants

Foodstuffs	**Common Adulterants**
Milk and milk products	Water, removal of butter and addition of refined oil.
Milk, liquid	Fat addition of skimmed milk reconstituted from skimmed milk powder.
Milk powder	Starch, dextrins.
Cream	Other fats.
Ice-cream	Non-permitted colour, artificial sweetness, other fats and gelling agents.
Butter	Other fats.
Ghee	Hydrogenated fat.
Vegetables oils and fats	
Vanaspati	Animal fat and other high melting fats.
Vegetable oil	Argemone oil, mineral oil, orthotricresyl phosphate, cheap non-edible oils.
Spices and condiments	
Whole turmeric	Coating of lead chromaie or coaltar dyes.
Turmeric powder	Coaltar colour, yellow earth, starch or talc coloured yellow by coaltar dye.
Curry powder	Starch coloured red by coaltar dye.
Coriander seeds	Other seeds coloured green.

Contd...

Foodstuffs	*Common Adulterants*
Coriander seed powder	Powdered bran or sawdust coloured green with dye.
Chilli powder	Starch coloured brown by coaltar dyes.
Mustard	Argemone seeds.
Cumin	Artificial jeera like product.
Black pepper	Dried papaya seeds.
Asafoetida	Resins and other plant gums.
Cereals	
Wheat and rice	Stones.
Wheat flour	Tapioca flour, talc
Semolina	Tapioca semolina.
Pulses	
Bengal gram dal	Khesari dal (Lathynis sativus)
Red gram dal	Coloured yellow with coaltar dye.
Bengal gram flour	Tapioca flour or starch coloured yellow with dye.
Sweetening agents and soft drinks	
Honey	Coloured canesugar syrup.
Soft drinks	Artificial sweeteners (Saccharin).
Beverages	
Coffee powder	Exhausted coffee powder, roasted husk or date seed or tamarind seed powder.

Contd...

Foodstuffs	***Common Adulterants***
Tea	Other leaves with added colour, exhausted tea leaves.
Miscellaneous	
Processed arecanut (Supari)	Other seeds or nuts broken and coloured.

Consumer Protection

For consumer protection there are a number of laboratories in central and state level. They are set up to collect samples and to analyse it. Some of the Government agencies are:

(1) Municipal Laboratories in big cities.

(2) Food and Drug Administration Laboratories of States.

(3) Central Food Testing Laboratory of the Government of India, and

(4) Laboratories of the Export Inspection Council.

Municipal Laboratories

In municipalities of big cities like Mumbai, Delhi, Chennai and Kolkata they have set up their own testing laboratories. According to the Prevention of Food Adulteration Act the Municipal analyst qualification must be scrutinised.

Chemical and microbiological analysis of all kinds of foods are essential in the laboratory. Nutrient analysis like fat content of milk, contamination of water or pesticide residue in parts per million in foods, etc. are some examples of the test conducted there.

These laboratories are under the Health Officer of the state. Health inspectors under him make periodical rounds of the various wards and collect samples of food suspected to be adulterated. Samples are sealed in the presence of the vendor and witness and sent to the municipal laboratory for analysis. The Health Officer can take action against the manufacturer of the food if he is found to be guilty.

The manufacturer can defend himself in the court of law. Depending on the severity of crime he can be punished by a fine or imprisonment or both.

Food and Drug Administration: Some state Governments have laboratories throughout states. The Director of the Administration is empowered through his staff to collect samples which are suspected to be adulterated and have them analysed by the Health Officer of a municipality. The Food and Drug Administration has jurisdiction all over the state as compared to municipal laboratories which work only within city limits.

The Central Food Testing Laboratory: The Government of India has established a Central Food Testing Laboratory in Kolkata for carrying out analysis of all foods. These laboratories are very well equipped and carry out even sophisticated and sensitive analyses.

These laboratories are the ultimate authority in determining whether a food sample is adulterated or not in cases of conflicting reports of analyses from two laboratories, the authorities concerned usually refer the case to this laboratory. Usually analysis carried out by these laboratories are accepted as the last word in analysis of food.

Export Inspection Council Laboratory: The Government of India has made it mandatory for all exporters to have their products analysed so as to ensure the quality of foods exported. Frozen food items like seafood, canned fruits and vegetables are also analysed before export. It also sees to it that all the foods exported conform to the minimum requirements as laid down by the council. The council has laboratories in all the major parts and the samples analysed. Without the council's certificate no food can be exported.

Quality Checking Agencies

Quality Control Laboratories of Companies: Most of the companies which manufacture food products have well equipped quaℓity control laboratories to check the quality of their products. The ingredients used for manufacturing and the products are examined physically, chemically and microbiologically. The taste, weight, volume, colour, keeping quality, packaging all these factors are checked.

In the dairy industry microbiological centre of the milk, are taken and the fat content determined, the former gives an idea of the sanitary condition under which the milk was collected, the latter whether any cream had been removed. Proper pasteurisation is also checked before supply.

In canning fruits and vegetables the fresh materials are visually inspected and all the bruised and spoiled ones removed. Oxidity of fruits like mangoes, cherries, tenderness of vegetables like green peas are determined to estimate duration of processing time, accelerated storage tests at higher temperature one conducted to ensure the keeping quality of the canned food.

In freezing, quality control checks at various stages of processing are carried out to see that the final frozen product conforms to the prescribed microbiological standards, weights after thawing and drying are taken to make sure that the customer gets the right weight. Biscuit, dairy, fruit and vegetable canning industries set up their quality.

Quality Control Laboratories of Consumer Cooperation: Apna Bazar in Mumbai, Super Bazar in Delhi, Chinthamani in Coimbatore have criteria for selection of food items which they sell in their shop. If the qualities are not fulfilled they will not accept materials.

Testing Laboratories: There are a number of private testing laboratories which carry out tests of all food materials on payment. Government has recognised certain private laboratories and certificates given by them are valid for legal purposes.

Consumer Guidance Society: A consumer guidance society has been formed in India with Bombay as its headquarters and with branches in major cities. They create consumer awareness of the various forms of adulteration and develop consumer resistance to such adulterated food products through various communications and mass media.

Importance of Dietary Food: Food is essential for human existence just like the air we breathe or the water we drink. The food that we eat is utilised in the body and the assimilated substances are used for the growth and maintenance of tissues. A living organism is the product of nutrition. The human being requires more than 45 different nutrients for its well-being. Food materials ingested by the body are digested, absorbed and metabolised. Useful chemical substances derived from food by the body are called nutrients.

A number of foodstuffs have to be selected to get all the nutrients. The health of a person depends on the type and quantity of foodstuff he chooses to make his diet For sustaining healthy and vigorous life, diet should be planned according to the principles of nutrition. Extensive research work carried out on human beings and on experimental animals throughout the world has provided us with sufficient knowledge on nutrition and health. World Health Organisation defines health as "the state of complete physical, mental and social well being and not merely the absence of disease or infirmity".

Importance of Food

Nutrition is the science of food values. It is relatively a new science which was evolved from chemistry and physiology. Nutrition is often mentioned as a branch of chemical science or biochemistry. The effect of food on our body is explained in nutrition. In other words, nutrition is defined as food at work in the body. In a broader sense nutrition is defined as the combination of process by which the living organism receives and utilises the materials necessary for the maintenance of its functions and for the growth and renewal of its components. Nutrients are defined as the constituents of food which help us to maintain our body functions, to grow and to protect our organs. There are six major nutrients in our body. They are carbohydrates, proteins, lipids, vitamins and minerals. The human body requires 17 vitamins and 24 mineral elements for various day-to-day activities. The composition of human body is 60-62 per cent water, 17 per cent proteins, 14 per cent fat, 6 per cent minerals and 1 per cent carbohydrates. In infants the percentage of water is more as compared to an adult. In women water content is slightly lower whereas fat content is more than in men. Fat deposition in the body increases with age.

Percentage Composition of Human Body

Nutrients	*Man*	*Woman*
Water	60-62	54
Protein	17	15
Fat	14	25
Minerals	6	5
Carbohydrates	1	1

Vitamins are present in negligible amounts.

Relationships of food to health have been made from the research conducted by chemists, microbiologists, pathologists and nutritionist from the past two centuries.

Human nutrition is governed by many factors like food habits and behaviour, food beliefs, ethnic influences, geographic influences, religious and sociological factors, psychological factors, food and production, income, national and international food policies, food technology, processing, fisheries, transportation, marketing, educational status and other mass media facilities.

The benefits of good nutrition are health, happiness, efficiency and longevity.

Sufficiency of Diet

When all the essential nutrients are present in a correct proportion as required by our body, it is called optimum nutrition or adequate nutrition. Optimum nutrition is required to maintain good health.

There are certain signs of good nutrition. They are height and weight for the age, clear complexion, fresh and lively skin and hair, healthy pink nails, correct posture and gait, inquisitive and alert eyes, good appetite and bowel evacuation, emotional maturity and confident deeds and pleasing personality and optimism in life and overall health.

Undernourishment: When almost all nutritions are below the requirement, the condition is known as undernourishment Under nourishment may be defined as a state of partial starvation. An undernourished person manifests symptoms of deficiencies and feels unwell. Poor body weight, poor resistance to infection, weakness, apathy and general ill-health are symptoms of undernourishment.

Malnutrition: Malnutrition is an impairment of health either from a deficiency or excess or imbalance of nutrients. Malnutrition is a condition when one or two nutrients are less or are in excess in the body. This again manifests in disorders and discomforts. Severe malnutrition in certain phases of life can do irreparable damage to the body. Physical, mental and intellectual well-being in a person is affected due to malnourishment. Emotional upset and intellectual dwarfism create personality problems in later life. Impaired functional ability and deficient structural integrity result from malnutrition.

Marasmus, kwashiorkor, xerophthalmia, scurvy, rickets, osteomalacia, beri-beri, pellagra and anaemia result from the deficiencies of protein, calories, vitamin A, vitamin C, vitamin D, B-complex vitamins and minerals, respectively. Malnutrition creates lasting effect on the growth and development of a person. Permanent retardation of the central nervous system occurs due to protein calorie malnutrition in early life.

Thus nutrients from food sources enable one to keep fit and maintain health. These substances include energy which gives the capacity to work, proteins which form our body muscles, bones, blood, body fluids, enzymes, hormones, antibodies, organs, skin, hair and nervous tissues. Food also supplies minerals and vitamins which protect our organs and regulate their functions and other physiological processes.

Classification of Foods: Based on these functions, foods are grouped into energy-yielding foods, body-building foods and protective foods. Nutrients which engage in these activities are known as energy-yielding nutrients, body-building nutrients and protective nutrients. Carbohydrase, fats and proteins release energy on metabolism in our body.

1 gm of carbohydrates provides	- 4 kcals
1 gm of proteins provides	- 4 kcals
1 gm of fat provides	- 9 kcals

Cereal grains like rice, wheat, ragi and maize, roots and tubers like potato, sweet potato and tapioca are good sources of carbohydrates. Fats are considered the concentrated source of energy as they supply more than double the quantity of calories compared to carbohydrates and proteins. Though proteins yield energy, normally they are considered body-building foods. Nutrients like proteins, mineral salts like calcium, phosphorus, iron and water are body-building nutrients. Their sources are body-building foods. Just like a house is built from wood, brick, cement, sand, iron rods, nails and other materials our organs in the body are built from the substances derived from food.

Protein foods like milk, meat, fish, eggs, pulses, grams and nuts are essential to build our tissues and to form blood.

Our body functions are regulated by water, certain minerals and vitamins. They are essential for the well-being and working of the body. They are called protective foods. Foods rich in protein, minerals, vitamins and water are termed as protective foods. Water is necessary for various body processes.

Minerals like calcium help in controlling blood clotting, muscular contraction and for the efficiency of heart muscles. Iron is essential for blood formation and iodine is necessary for regulating body function through the thyroid gland.

Vitamins are essential for regulating the body process such as growth, muscular coordination of various organs and functions of several organs like eyes, ears, nose and skin.

Thus foods play a prominent role in providing physical, mental and social well-being which is otherwise known as health to the people. Health is reflected in a person's nutritional status. Nutritional status is the condition of the individual as influenced by the utilisation of the nutrients. Dietary history, physical examination and laboratory examinations reveal it

Consumption of Diet

Food industry has brought out some novel protein-rich foods which can be used for combating malnutrition. Some of them are from unusual food sources which were never considered as consumable by our forefathers. Leaf and grass proteins, food yeast, algae and microbial synthesis of proteins from hydrocarbons are some of the novel food sources now available.

Leaf Protein Concentrates: Proteins from fibrous materials of leaf are separated and leaf protein concentrates with 30 to 35 per cent of the original nitrogen can be extracted from leaves. Outstanding work in this field has come from Pine et al in 1961. Leaf protein concentrates can be incorporated with milk or in common flours or dal powders for children's foods.

Food Yeast: Yeast is capable of converting carbohydrate to protein. About 16 to 33 per cent of this type of conversion is possible by yeast. Thus it can be incorporated at 5 to 10 per cent level in children's foods. Food yeast is a pale yellow product. It has a characteristic nut-like taste. This can be incorporated in low cost foods for infants, it can also be used for overcoming protein malnutrition among children.

Algae: A unicellular green algae chlorella contains a protein content of about 50 to 60 per cent. This can be incorporated with soups, noodles and in baked items. Even though it is nutritionally good, its acceptability is very poor.

Microbial Synthesis of Proteins from Hydrocarbons: Certain petroleum hydrocarbons can be converted to proteins through microbial use. Its utilisation for human beings is not established even though as an animal feed it is used.

Food Preservation

Modern technology has made many food sources into processed items. Food industry is developing in accordance with modern living. Almost all food groups are now processed into readymade or ready for table items. Pre-cooked or half cooked items are also available. Examples of some of these items are given below.

Cereals: Starch is a polysaccharide and it is used in cookery. It is manufactured from corn, potato, sweet potato and tapioca. In the market it is available as tapioca starch, arrowroot starch and starch products like sago or custard powder where cornstarch is flavoured with vanilla and edible colour and corn syrup.

Rice is available in processed form as instant rice or quick cooking rice. It is prepared by cooking rice and dehydrating the cooked rice so as to retain a porous structure.

Puffed rice from paddy and from parboiled rice is available in the market.

Cornflakes from corn can be used as a breakfast item.

Barley is processed into barley flour and pearl barley. Malted barley is breakfast item.

Oats is available as oatflakes in processed forms.

Breakfast cereals and infant foods are often prepared from cereals. Flaked items from all cereals, puffed wheat, rice, barley and corn are available in the market. Shredded breakfast cereals like shredded wheat is a pre-cooked item.

Children's Food

Most of the infant foods contain both cereals and milk and these pre-cooked cereals are eaten as gruels with water or milk. Rice noodles are popular in some of the Far East countries. Sago or sabudana is made from rice, or other cereals. Biscuits and breads are also processed cereal products.

Pulses are processed into puffed grams. Now commercial instant mixes of cereal, dal, fermented products are also available. Ready mixes of South Indian dishes like idli and dosa are prepared by using baker's yeast or baking soda and acidifying agents like citric acid.

Puffed chickpeas and peas and canned dry peas are available as processed items.

Roasted groundnut and protein isolates from soyabean or peanut can be used in the production of toned milk and protein-rich biscuits or bread. Soyabean milk powder and butter are available in the market.

Milk powders, malted milk powders, various weaning foods, ice-cream mix, cream and condensed milk, cheese, khoa-based sweets, channa or paneer-based sweets like rasogulla and sandesh are some of the processed milk products. New milk-derived products like Miltone and Chai-sathi are available, especially in North India.

Fruits are processed into fruit juices. Today even bananas and guavas are used to prepare juice. Fruit juice concentrates and fruit juice powders are made from juices along with syrups or squashes. Cordials, ketchups, sauces, jellies, jams, marmalades, artificial syrups, dried fruit and salad, preserves, candies, glaced and crystallised fruits and fruit-cereal flakes are available in the market as processed foods.

Vegetables are processed into soups, purees, canned or frozen and dehydrated forms.

Egg powder, egg albumin powder, custard or caramelised custard powder and readymade puddings are some of the processed forms of egg available in the market.

Meat is processed into canned meat, cured meat, smoked meat, sausages, cutlets, soup powder and as dehydrated meat.

Strained baby foods with meat are being manufactured in some places.

Fish is processed into frozen, canned and dried forms. Fish protein concentrates and fishmeal and strained baby foods with fish are also available as processed items in developed countries.

All food items are processed in very many ways and they are marketed in various forms. Pickled and fermented items of innumerable sources are also available in the market.

Everyone knows that good diet is needed to sustain good health. By health we mean the well-being of an individual in physical, emotional and social conditions. Just the presence or absence of a disease cannot label a person as healthy. Sound emotional and mental condition of an individual is contributed by good nutrition. A healthy person will have a positive attitude towards life. A good natured person full of life reflects his better health standards. A good or adequate diet is known as 'balanced diet'. A balanced diet yields daily nutrients in the proper amounts and proportion required by the body.

Necessity of Nutritional Diet: Nutritional requirements vary according to age, sex, physical activities and other physiological conditions. The National Institute of Nutrition is engaged in research work on the dietary pattern of Indians and their requirements of various nutrients. The Nutrition Advisory Committee of the Indian Council of Medical Research made recommendations in 1944 for all nutrients.

They were revised in 1958 and 1968. In 1980 further recommendations were made by the Nutrition Advisory Committee and the latest requirements are available now for formulating balanced diets for various groups. A balanced diet must supply enough food to the body for deriving energy. Energy is simply expressed as the capacity to do work. For basic activities like heart beat, breathing and excretion energy is required.

Energy is stored as carbohydrate in foods. Cereals like rice, wheat, millets like ragi or bajra contain much carbohydrates in the form of starch. Starchyuits like banana and roots like tapioca, potato, yam, colocasia and sweet potato are rich in carbohydrates. Energy requirements for various age groups are given elsewhere and for an adult sedentary man 2,400 kcals are required and for a sedentary woman 1,900 kcals are recommended.

In a balanced diet energy from cereals should not exceed more than 75 per cent of total requirements. Energy derived from fat or oil should not exceed 15 per cent of the total calories and from refined carbohydrates (sugar and jaggery) has been recommended around 5 per cent of the total calories.

A balanced diet must supply enough protein for building

up our tissues in various parts of our body. When some tissues get worn out or used up proteins repair and replace them. Proteins are essential for the secretion of digestive juices and synthesis of enzymes and hormone production.

Dietary proteins are used again for the synthesis of body ins. Protein requirements must be met by dietary proteins. Usually in requirement is in terms of body weight, that is, 1 gm of protein to body weight is the recommendation.

Even though animal ins are of superior quality it is not a practical suggestion to depend on them for dietary proteins. In an average Indian diet pulses are a very important source of proteins. When nutritional scores are given to jus proteins egg has a high score of 90 per cent whereas vegetable proteins score only 45 to 50 per cent in the case of pulses and 45 per cent to cereals.

But it can be improved by incorporating pulse and cereal proteins; 40 to 60 gms of pulses and 150 to 250 ml of milk can provide enough proteins in a vegetarian diet For non-vegetarians 20 to 30 gms of pulses and one egg or 30 gms of meat or fish and 100 to 150 ml of k provide the protein requirement If no pulse is used non-elarians can take two eggs or 50 gms of meat or fish or 30 gms of fish neat and one egg.

Fats and oils are other major nutrients in a balanced diet. Fats are concentrated forms of energy. Fat soluble vitamins are utilised only in presence of fat For males 40 to 65 gms of fat and for females 20 to gms of fat must be included in a balanced diet. Vitamins are essential the proper functioning of our body. Small quantities of vitamins own dramatic effect on our body.

Organs like eyes, alimentary canal, in and mucous membraneous structures depend on vitamins for their formal structure and functions. Leafy vegetables, other vegetables, yellow vegetables and roots and tubers provide almost all the vitamins required by our body. About a quarter kilogramme of vegetables is required in our daily diet to ensure vitamin supply.

Raw vegetables or sprouted grams or fruits are essential to supply vitamin C. Minerals like calcium, phosphorus, iron and trace elements and salts are essential to maintain good health. Sodium chloride is essential to regulate water metabolism in our body.

A balanced diet provides enough vitamins and minerals from natural sources. Tonics or supplements are not required or people who consume balanced diets. The Indian Council of Medical Research has designed Recommended Dietary Allowances for various nutrients.

They have also formulated composition of balanced diet for various groups. These allowances are designed to serve as a guideline for planning diets. Suggestions are given for substituting foods for-non-vegetarians and additional allowances during special conditions.

To include the above nutrients in daily diet certain guidelines are suggested by the ICMR. Food items which contribute the same nutrients are grouped together and certain food groups are evolved. Foods are grouped into 11 groups based on their nutritive value. They are cereals and millets, pulses, nuts and oilseeds, vegetables, fruits, milk and milk products, meat, fish, eggs and poultry, fats and oils, sugar and jaggery and spices and condiments.

Cereals and millets constitute 70 to 80 per cent of calories and proteins of low income group. Except ragi all cereals are poor in calcium. Dried pulses are rich in proteins and a cereal-pulse combination is an excellent substitute for animal proteins—pulses are also good sources of many B vitamins and sprouting enhances the vitamin C content of pulses. Nuts and oilseeds are good sources of proteins, certain B vitamins, vitamin E and minerals like phosphorus andiron. Vegetables are the storehouse of carotene, riboflavin, folic acid, vitamin C and calcium.

Vegetables also supply water and roughage to the body. Leafy vegetables, other vegetables and roots and tubers are the different types of vegetables. Fruits in general are a good source of vitamin C. Milk is nature's best food as it is almost complete except for iron and vitamin 'C.

Like milk, egg also contains proteins of high biological value. Egg is a rich source of vitamin A and some B vitamins. Animal foods like meat, fish and poultry are rich sources of proteins and B vitamins.

Fatty fish are good in fat, vitamin A and D and small fish with bones are good sources of calcium. Liver and other organ meats are good sources of proteins, vitamin A, B complex, B_{12} and iron.

Sugar and jaggery are good sources of energy and honey and jaggery contain minerals like iron. Condiments and spices are mainly used to enhance palatability.

For non-vegetarians pulses can be reduced to 50 per cent and instead of that, one egg or 30 gms of meat or fish can be included. If no pulse is taken two eggs or 50 gms of meat or fish can be included.

Significance of Balanced Diet: For planning a balanced diet for labourers, low cost nutritious foods must be included. Ragi, tapioca, sweet potato, clusterbeans, bittergourd, soyabeans, drumstick leaves, various types of leafy vegetables, cowgram, horsegram, groundnut, Indian gooseberry, guava, papaya, mango, cashew fruit and other seasonal fruits can be used liberally to make a low-cost balanced diet Sprouted pulses and cheap vegetables that can be used as raw salads also must be included to meet the vitamin C requirement.

The ICMR Advisory Committee (1981) recommended the composition of balanced diet and it is given below:

Food Item	*Adult Man* Sedentary	*Adult Man* Moderate Work	*Adult Man* Heavy Work	*Adult Woman* Sedentary	*Adult Woman* Moderate Work	*Adult Woman* Heavy Work
Cereals	460 gm	520 gm	670 gm	410 gm	440 gm	575 gm
Pulses	40 gm	50 gm	60 gm	40 gm	45 gm	50 gm
leafy vegetables	40 gm	40 gm	40 gm	100 gm	100 gm	100 gm
Other vegetables	60 gm	70 gm	80 gm	40 gm	40 gm	100 gm
Roots and tubers	50 gm	60 gm	80 gm	50 gm	50 gm	100 gm
Milk and milk products	150 ml	200 ml	250 ml	100 ml	150 ml	200 ml
Oil and fat	40 gm	45 gm	65 gm	20 gm	25gm	40 gm
Fruits	60 gm	60 gm	60 gm	60 gm	60 gm	60 gm
Sugar and jaggery	30 gm	35 gm	55 gm	20 gm	20 gm	40 gm

For non-vegetarians pulses 50 per cent less + one egg or 30 gms fish or meat, if no pulse, 2 eggs or 50 gm fish or meat.

A Sample Diet for a Sedentary Woman

Time	*Meal*	*Menu*
6.30 a.m.	- Bed coffee	
8 a.m.	- Breakfast	- Chapathi, Groundnuts-Tomato curry, Coffee
12.30 Noon	- Lunch	- Rice or Chapathi, Brinjal and horsegram curry Amaranth or spinach pugath, Lassi, Carrot salad.
4 p.m.	- Tea	- Vegetable vermicelli uppuma, Mint leaf chutney, Tea.

Contd...

Time		*Meal*		*Menu*
8 p.m.	-	Dinner	-	Wheat dosa or phulkas, Sprouted green gram curry, Fruit cup.

Quantities of various food groups can be distributed according to the composition given by ICMR. This above diet provides:

Calories - 2420 kcals

Proteins - 53 gms

The idea that the body's needs are definite and can be measured is a very recent one. We are so accustomed to eating to satisfy hunger that for centuries this criterion accepted by most people without question. Plenty was, in general, associated with health and want with starvation.

And yet, throughout the ages there have been those few who have seen a little further into the meaning of nutrition. We know from the book of Daniel of how he asked that he and his companions might eat a simple diet of pulses and water and how, after ten days during which this was permitted, 'their countenances appeared fairer and fatter in flesh than all the children which did eat the portion of the king's meat.' This was probably the first dietary experiment ever to be recorded.

We also know how, in the Middle Ages, illness and loss of life from scurvy led to the recognition that food could prevent and cure certain diseases but the connection between man's daily food and his health was only dimly acknowledged and in a very limited field.

Fortunately, the pleasures of a mixed diet led those who could afford it to eat a variety of foods. The greater the variety, the greater chance there was that all the vital and necessary factors would be included. Hence disease was most prevalent in institutions where the food was restricted and monotonous and among the poor who could not afford variety.

It was Sanctorius (1561-1636) who showed, by sitting on a chair suspended from the arm of a balance while he ate a meal, that afterwards the increase in weight which had taken place was gradually lost again. He described the loss as due to 'insensible perspiration.'

This proved the first basic fact that body-weight was connected with food eaten and that living, even without any conscious exertion, causes loss of weight.

This led in time to the acknowledgment that all the activities of the body derived their energy from the slow combustion of the foods eaten or (in starvation) of the body itself.

An adequate diet is one of that provides all the essential nutrients in sufficient quantities to meet the needs of the body. Thus far, the energy and nutrient needs, including carbohydrates, fats, proteins, minerals, vitamins, and water. The information concerning each of the nutrients in terms of the day's food is now brought together.

Easy rule-of-thumb methods for knowing which foods and how much of each one are needed to provide an adequate diet have been developed on the basis of *food groups;* for this purpose, foods similar in nutritive value are considered together as a group. The pattern for the adequate diet is developed by recommending sufficient food servings from the various groups.

Various Human Requirements: The food shortages during and after the 1914-18 war and the resulting ill-health of children, especially in Eastern Europe, made the general public realise for the first time that starvation was not the only consequence of a poor diet. Enough food to satisfy hunger sometimes failed to protect from diseases of malnutrition.

Food had to be of the right and, to be fully protective, had to contain the essential vitamins as well as the basic nutrients: proteins, fats, carbohydrates and mineral salts.

Information to have extracted from old records about diets known to have been associated with deficiency diseases. Beriberi had occurred most commonly in countries where rice formed the bulk of the food eaten and had been known to disappear when some of the rice was replaced by vegetables.

A diet provided for the U.S.A. garrison of Porto Rico consisted of rice, beans, potatoes, onions and meat with some extra fruit and vegetables, and on this, cases of beriberi occurred. Using recent values for the vitamins B_1 content of these foods, we find that the daily diet contained about 0.5 mg. of vitamin B_1 per man. This represents an average intake, and implies the some men ate less and some more, the former being without doubt those who succumbed to beriberi. The rations were later improved.

Less rice and onions were given and more beans, potatoes and meat, and with the change, beriberi disappeared from the camp. The vitamin B_1 of the improved ration calculated by modern methods was just over 0.8 mg. per man — an increase of only 0.3 mg. but it made the difference between beriberi and protection. This difference is about two-millions of the total weight of a man's daily food, so it is not surprising that vitamins, active in such minute amounts, remained undiscovered for so long. During the 19th century, scurvy was common in prisons and in other institutions, not to be punitive. In 1822 the meals in Millbank Penitentiary are described as:

For females — morning 9 oz. bread

— pint gruel (made with barley)

noon 9 oz. bread — pint soup evening — pint soup.

The bread would have been made from stone-ground flour and would contain plenty of vitamin B_1, so as we should expect, there is a history of beriberi in such a prison, but any vegetables used in the preparation of the soup would be unlikely to retain any vitamin C during the cooking and it recorded that 448 prisoners did in fact show signs of scurvy in one year.

During the war of 1914-18, both scurvy and beriberi occurred. Soldiers of the Indian Army who ate few vegetables and little if any meat, suffered from scurvy. The British on the other hand, who ate vegetables, but while in India had rice or refined cereals, went down with beriberi. In 1916, special hospitals for the admission of cases of scurvy were operated in Baghdad and other parts of the Middle East.

A valuable study was made between 1920 and 1930 on the effect of increasing the quantities of protective foods in the meals of schoolboys. Records were kept of the heights and weights of boys between the ages of 10 and 18 at a large public school. Conditions of housing, the amount of physical exercise taken and other possibly influential factors remained relatively constant during period and the boys were, by accepted standards of nutrition, well fed.

However as a result of an increasing awareness of the importance of vitamin-containing food the authorities altered the diet, so that the boys ate less bread and more milk, vegetables and butter. The results of these changes as seen in records kept over a long period were published by the medical officer (Friend, 1935).

At all ages there was an increase in both height and weight, this being greatest, as one would expect, during the years of most rapid growth.

It may be argued that increase in size does not necessarily mean better health and we are all familiar with overweight children who are abnormally fat but, in general, and where large numbers are concerned, the rate of growth of children can be taken as one indication of physical progress.

This study as important a being the first of its kind undertaken on an adequate scale and over a long enough period to give reliable results.

With the increasing consumption of vegetables and fruits in Great Britain, scurvy, at one time described as the 'English disease,' became rare and the introduction of cod liver oil as a supplement for infants reduced the incidence of rickets, but the new knowledge of the importance of vitamins as protective factors roused great interest in nutrition and people began to wonder how present-day diets would stand up tests of adequacy.

The first report of a large-scale investigation into modern diets was published by J. Boyd Orr in 1937. He divided the community into six sections according to income and estimated the amounts of money spent on food and also the consumption of protective foods in the different income groups. He found that with rise of income, more protective foods were eaten and that in three lowest groups, the amounts of Vitamins A and C eaten were very low.

Investigations on similar lines were made in the United States and from all the data collected, it became possible to make a fairly good estimate of the amounts of the more common vitamins contained in the daily food of normal healthy people and also the amounts below which deficiency diseases were liable to occur.

Tables setting out probable human requirements were published by a Technical Commission of the League of Nations in 1937 and by the Government Bureau of Home Economics, U.S.A. (Stiebeling, 1933). This was before the adoption in the United States of international units, so the results are not comparable but the attempt to express human needs in terms of quantities of vitamins was in itself a great advance.

Later, tables using international units were prepared by the National Research Council of the U.S.A. and these gave (except for vitamin C) figures surprisingly close to the League of Nations recommendations. The most recent publication is that the British Medical Association, 1950.

	League of Nations 1937	*National Research Council 1945*	*British Medical Assoc, 1950*
Vitamin A	2,000-4,000 i.u.	5,000 (optimum)	i.u. 3,000 i.u.
Vitamin B_1	1 mg. approx.	1.5 mg.	1.4 mg.
Vitamin C	30 mg	75 mg.	20 mg.
Vitamin D	—	—	400 i.u.

During the war of 1939-45 there were many cases of dietary deficiency including beriberi in the Japanese prison camps. Some of the doctors who were among the prisoners kept record of the occurrence of beriberi and also of the vitamin B_1 of the diet and the history of the Porto Rico garrison was repeated.

When the intake fell blow 0.5 mg. vitamin B_1 (or about 0.2 mg. per 1,000 nonfat Calories) men succumbed to beriberi; when it rose above these levels, the men's health improved (Smith and others, 1951).

The Register-General's Statistical Review which is published every year gives details of the number of deaths from different diseases. Although the causes of scurvy and rickets were known in the 1920s, and the foods effective for prevention were available, the diseases did not disappear immediately. There was, however, a steady drop in incidence. This demonstrates that not only must protective foods be available, but the public must be educated to use them properly.

Free supplements for the provision of vitamins A, C and D were made available for infants during the war years but all mothers did not avail themselves of the opportunity to have them. The number taking them has increased considerably in the last few years with a corresponding drop in rickets almost to the point of disappearance (Registrar-General, 1949, 1950 and 1952) and the health of children in this country has steadily improved.

Different Problems

Most of the vitamins, being growth-promoting substances in the sense that they are needed for full structural and functional development are needed greater amounts during childhood when growth is taking place and during pregnancy and lactation for the same reason.

Vitamin A: Vitamin A is stored in the liver and during adult life the daily needs can be supplied to a considerable extent from these reserves. Volunteers who deliberately consumed diets very low in vitamin A were able to continue doing so for several months without developing any gross signs of deficiency. But in spite of this there is general agreement that it is better to consume enough to guard against any chances of deficiency and, in the table, the recommended levels of vitamin A vary very little.

Whether taken as vitamin A itself or in the plant form carotene, this factor is always converted into vitamin A used as such in the animal body. But the two forms are not of equal physiological value, unit for unit. It is usually reckoned that the number of international units needed in the form of carotene is about twice the number needed as vitamin A itself. Even this ratio is not the same for all foods.

The carotene of green vegetables supplies to the body about half as many units of vitamin A as the number obtained by chemical estimation, but yellow vegetables like carrots supply only about one-fifth or one-quarter (Graves, 1942).

Increased Needs for the Vitamin A: Vitamin A being used by the body to maintain the health of the skin, teeth and mucous linings of the respiratory organs is needed especially in illness in which these are involved. To guard against them, supplies in the daily diet should be kept up to the standards given. In treatment higher doses are available in pharmaceutical preparations can be prescribed. Dases up to 100,000 or 200,000 i.u. are sometimes given daily.

The B Vitamins: The need for vitamin B_1 varies according to the amount of· food eaten and the energy expended in movements in all kinds.

As muscular activity, which depends on the oxidation of carbohydrates, continues throughout life, vitamin B_1 must be continuously supplied. Moreover, this vitamin differs from vitamin A is not being stored for long periods. In general, the fat-soluble vitamins tend to be stored in the body, but the water-soluble vitamins, *i.e.* those of the B group and vitamin C, are not stored. If more is eaten than needed it is rapidly excreted again in the urine.

Because vitamin B_1 is used day by day, as it is consumed, it has been much easier to find out exactly what the daily needs are in fact, as early as 1934, Cowgill was able to suggest a mathematical formula to express the daily needs in terms of body weight and calorific intake.

Increased Needs for Vitamin B: As vitamin B_1 is concerned with carbohydrate metabolism, any conditions which involve a higher calorific intake, e.g heavy manual work, climbing, Article exploration, naturally call for higher intake.

In those diseases, too, in which the rate of metabolism is increased, more vitamin B_1 is needed.

Riboflavin and nicotinic acid also take part with vitamin B_1 in energy relations and the amounts needed are related to the amount of vitamin B_1 In foods, these vitamins of the B complex occur together and a shortage of one usually implies a shortage of the others.

It is only when the staple cereal happens to be grossly short of a particular factor as, for example, polished rice and vitamin B_1 or maize and nicotinic acid, that well-defined deficiency diseases such as beriberi and pellagra occur. In countries where mixed diets are customary, deficiencies are usually less severe but of a mixed type.

For example, a mild shortage of riboflavin will produce, in some individuals, roughness and redness of the skin over the nose and some cracking of the lips, easily cured by the consumption of extra quantities of one or two foods rich in riboflavin such as eggs or liver.

Similarly a shortage of nicotinic acid, not enough to cause pellagra, may produce digestive disturbances, a reddening of the skin of the face and hands and feelings of depression. Nicotinic acid is concerned in healthy people with the control of movements of precision and if its supply is limited subjects may find themselves unusually clumsy.

The varying responses of individuals to diets deficient in the B vitamins was puzzling until, as a result of a long series of observations, it was learnt that these factors are also synthesised within the body of man and the higher animals though not by it.

The intestinal tract contains warm semifluid material rich in nutrients and is, as we know, a medium in which there is normally an expensive bacterial flora. These organisms are not disease-producing. They merely live in and with man. Their power of synthesising vitamins is a relatively recent discovery. They seem to specialise in the factors of the B group but some can also synthesise vitamin K (Najjar and Barrett, 1945).

Ruminants, whose large rumen in the first part of the digestive tract is the site of vitamin formation, obtain so much riboflavin and nicotinic acid from bacterial activity that they fail to develop signs of deficiency even on diets lacking these vitamins. But in man the bacteria carry on their synthetic activities much further down the digestive tract and, although undoubtedly some of the B vitamins produced in this way are available for human use, enough of each is not formed to satisfy all human needs.

It is, however easy to see that in conditions such as diarrhoea in which the contents of the alimentary tract move along quickly, less than the normal amounts may be absorbed and symptoms of deficiency may occur if the condition is not checked. The recent use of some of the new drugs, sulphonamides and antibiotics has also revealed that these, too, affect the activities of the vitamin-B-producing bacteria.

The drugs inhibit not only the growth of disease-producing organisms but also the synthesis of the B vitamins in the intestine. This interference with production may be great enough to cause true secondary deficiencies in patients who may not have been very well supplied with vitamins before. Cases have been observed with signs of peripheral neuritis, enlarged heart and swelling of the limbs, indistinguishable from true beriberi. Other patients develop a sore mouth and red scaling of the skin recalling deficiencies of nicotinic acid and riboflavin.

In the treatment of these secondary deficiencies very large doses have to be given — a hundred or more times the amount needed for normal health.

Vitamin C: Vitamin C, also water-soluble, is another vitamin which should be taken daily, although the body stores are not exhausted as rapidly as those of vitamin B_1

There is difference of opinion about the amount of vitamin C needed to maintain health. The American in which estimate is higher than the British. An experiment in which volunteers did without any vitamin C for months showed that they could do so without developing any gross signs of scurvy, and the quantity needed for health has consequently been placed at a lower level than formerly. But, as the signs of deficiency develop in such an insidious manner, it is probable that minor degrees of vitamin C deficiency are fairly common though not always recognised as such.

It is known now that vitamin C is concerned not only in protection against scurvy but in the healing of wounds, in the prevention of shock from injury, in the tolerance of high atmospheric temperature and in the control of body fluids. These processes are common in everyday life and probably account for the daily use by the body of vitamin C.

Increased Needs for Vitamin C: When any of these, conditions are actually present or exaggerated the need is correspondingly increased. Very large doses (up to 1 g.) are given before operation to prevent shock and in the treatment of severe injuries and larger than normal doses are given to man doing heavy work in an excessively hot humid atmosphere

Large amounts of vitamin C are available in fruits and vegetables and some reason or other an individual does not eat or can not eat these foods, then tablets of the synthetic vitamin or some other pharmaceutical preparation must be taken. The massive doses are, of course, always given by means of a pharmaceutical preparation.

Vitamin D: It will be noticed that the table gives the same figure for the human requirement for vitamin D except in the case of infants under one year and for expectant and nursing mothers. This uniform level is an fact a confession that very little is known about varying needs.

At one time it was assumed that vitamin D was needed only during the period of active growth. This view is no longer held and it is thought that the fragile nature of the bones people of older may be related to their low consumption of vitamin D foods.

The figures for human needs given in the B.M.A. table have been obtained by studies of dietary surveys, of individual diets, of diets associated with deficiency diseases and so on. The first estimates were made by the Health Committee of the League of Nations; these were followed by American estimates and finally the table quoted here was produced.

Considering the magnitude and difficulty of the task, it is astonishing how closely the various estimates agree and it is probable that these recently published figures are as near the truth as we are likely to get.

The question has to be asked: to what extent do individuals eat diets which satisfy these needs? How easy or difficult is it to obtain an adequate diet? There are several ways of approaching this problem. We can see whether there is any evidence that people are better fed now than formerly or we can study foods which are normally eaten in ordinary homes and see if they supply the needs.

Reference has been made to the effect of adding protective foods to the diet of schoolboys and to the increases in height and weight produced as a result. Data on the height and weights of London school children during the first half of this century have recently been published and these show the same kind of increase. It may not be totally due to nutrition, but it is not unreasonable to conclude that improved feeding due to increased knowledge of nutrition has also played a part in producing these effects.

If we approach the problem from the dietary angle, we can analyse representative diets and estimate their vitamin content. This is a useful exercise and a comparison between the foods eaten on several different days or by different individuals shows how wide the variation in vitamin content can be.

Following examples of meals will illustrate this point. These show that the value of a diet depends on the *choice* of foods, and that likes and dislikes are of themselves not reliable guides; there must be knowledge also.

Amount *Oz. i.u.*	*Vit. A or equiv. in carotene mg.*	*Vit.* B_1 *mg.*	*Ribo-flavin mg.*	*Niacin mg.*	*Vit C i.u.*	*Vit D*	
Meal 1							
Roast beef	3	42	33	195	3,900	—	
Yorkshire pudding							
3 oz. = flour	1	–	34	23	450	-	-
milk	1½	45	18	54	-	-	0.45
egg	½	125	19	85	-	-	8.5
cooking fat	1/8	-	-	-	-	-	-
Roast potatoes	6	-	180	26	1,200	15	-
Cabbage	4	288*	4	28	320	28	-
Apple pie= flour	½	-	171	11	225	-	
cooking fat	¼	-	-	-	-	-	
apple	2	4	6	-	50	1.6	-
sugar	¼	-	-	-		-	-
Custard= milk	2	60	24	72	-	-	0.6
powder	mg.	-	-	-	-	-	-
		564	335	564	6,145	44.6	9.55
		i.u.	mg.=	mg.=	mg.	= mg	i.u.
			.335	.564	6.145		
			mg.	mg.	mg.		
Proportion of day's need approx.	1/8	¼	1/3	½	All 1/40		
Meal 2							
Cheese	1	369	9	140	100	-	4
Lettuce	2	727*	34	40	200	6	-
Tomato	3	540*	60	30	300	21	-
Wholemeal bread	3	-	270	150	1,800	-	-
Butter	½	568	-		-	-	8.5
Orange	3	48	60	30	300	48	-
Coffee= milk							
and sugar	5	150	60	180	-	1	1.5
		2,402	493	570	2,700	76	14
		i.u.	µg.=	µg.=	µg.=	mg.	i .u.
			.493	.57	2.7		
			mg.	mg.	mg.		

Proportion of day's need approx		¾	1/3	1/3	1/5	All	1/90
Meal 3							
Fried white fish	4	-	40	160	2,400	-	-
Fried potatoes	6	-	180	60	1,200	12	-
Rice pudding							
4 oz.							
= rice	1/3	-	5	6	100	-	-
milk	2½	75	30	90	-	-	.75
sugar	1/6	-	-	-	-	-	-
Stewed apple, 4							
=raw	3	7	9	-	75	2.4	-
		82	264	316	3,775	14.4	.75
		i.u.	μg.=	μg.=	μg.=	mg.=	i.u.
			.264	.316	3.775		
			mg.	mg.	mg.		
Proportion of day's need approx		1/40	1/5	1/6	1/4	1/2	negli-gible
Meal 4							
Sausages, pork	3	-	210	150	1,500	-	-
Baked beans	3	35	60	30	600	-	-
Mashed potatoes	6	-	144	96	1,350	12	-
Prunes	4	103	32	68	-	-	-
Custard= milk							
powder	3	90	36	108	-	-	.9
		228	482	452	3,450	12	.9
		i.u.	μg.=	μg.=	μg.	= mg.	i.u.
			.482	4.52	3.45		
			mg.	mg.	mg.		
Proportion of day's need approx		1/13	1/3	1/4	1/4	3/5	negli-gible

For green vegetables: Carotene = 40 per cent as vitamin A

Yellow veg.= 25 per cent

Half the value of uncooked flour.

Where no value is available for vitamin B_1 in cooked food, the figure for the uncooked food is halved.

Universal Steps

Vitamins A and D are usually expressed in international units (i.u.)- Human or animal requirements for other vitamins are usually given in milligrams, but the value in foods is often expressed in microgrammes (1/1000s of miligram) so as to avoid the use of decimals.

In the tables, the value of *individual* foods are given in micrograms for the same reason but the *total* values and the requirements are given in milligrammes for comparison with other published figures.

Conversions —Weight and Household Measures

5 oz. milk	= 1 teacupful
1 oz. cheese is the same size as	1 oz. butter (1/4 of 1/4- 1b. pack)
1 oz. bread	= ¼ in. slice from square loaf
1 oz. cake	= portion 3 x 1 x 1 /2 in.
1 oz. Castle pudding is the same size as	1 oz. cake
1 oz. biscuits	= 2 biscuits (2 1 /4 in. in diameter)
1 oz. flour, rice, oatmeal, custard, powder, sugar, etc.	= 1 well heaped tablespoonful
½ oz.	= 1 level tablespoonful or 1 well heaped teaspoonful
1 oz. jam	= 1 desert spoonful
2 oz. potato (mashed), green vege. tables, root vegetables or peas	= 1 heaped tablespoonful
2 oz. tomato	= 1 egg-size tomato
4 oz. stewed fruit	= 5 small plumps (with juice) or 4 tablespoonfuls of soft fruit
3 oz. meat	= 4 thin slices (approx. 3 x 4 in.)
1 oz. bacon	= 1 rasher (approx. 1/8 x 7 x 2 in.)
4 oz. fish	= portion approx. 3 x 3 x 1/2 in.
2 oz. herring	= 1 small herring (approx.5 in. long)

Sufficient Diet

A plan for selecting an adequate diet developed around 4 *food groups* will now be discussed. There are other plans, developed on the same principle as the 4-group plan, differing only in number of groups.

All of the plans are developed in accordance with food habits of the people and the foods available to them. In case the habits of eating and the food supply are different, the food patterns are of necessity also different.

The 4-group plan is flexible, permitting for a wide choice of foods within most of the groups. Meals planned in one home are likely to differ from those in another, even though the meals in each home follow the same food pattern. Foods selected by persons eating out may vary widely, and still at the same time each is in complete accordance with the recommendations of the same food pattern.

Four-Food Group Plan: The 4-group plan (milk, meat, vegetable-fruit, cereal) was developed in 1958 and is a simplified version of the 7-group plan that originated during wartime 1943. In this 4-group plan the citrus fruits, green and yellow vegetables, and other vegetables and fruits are combined into one group, and the butter fortified margarine group omitted.

The daily quantities of each group of recommended to provide the *foundation* of an adequate diet and the main nutritive contribution of each group are given. Calories are the least well supplied by this pattern.

However, as a usual practice, facts and sugars, which are not listed, are eaten as such or combined in baked goods and desserts, with a result that the energy intake is increased.

Also to satisfy the appetite, additional servings from the food groups may be exported, which will contribute both to the energy and nutrient intake.

The 7 food groups	*Daily amounts*	*Min nutritive contributions*
Green and yellow vegetables	1 or more servings	Vitamin A value Ascorbic acid Iron
Oranges, grapefruit, tomatoes or raw cabbage or salad greens	1 serving	Ascorbic acid
Potatoes, other vegetables and fruits	2 or more servings	Vitamins and minerals in general Cellulose
Milk and milk products	Children ¾ to 1 quart Adults 1 pint	Calcium Riboflavin Protein Phosphorus
Meat, poultry, fish and eggs or poultry 1 egg (at	1 serving meat, fish Iron B- Vitamins least 4 a week)	Protein Phosphorus
Bread, flour and cereal (whole grain enriched or restored)	3 or more servings	Thiamine Niacin Riboflavin Iron Carbohydrate Cellulose
Butter or fortified margarine	2-3 tbsp.	Vitamin A Fat

The basic 7 food groups.

The 4 food groups	*Daily amounts*	*Main nutritive contributions*
Milk group: Milk, cheese, rice cream (cheese and ice cream can replace part of the milk)	Children 3-4 cups Teenagers 4 or more cups Adults 2 or more cups	Calcium Riboflavin
	Pregnant 4 or more cups	Protein
Women Nursing 6 or more cups Mothers	Phosphorus	
Meal group	Protein	
Beef veal pork, lemon,		Phosphorus
Poultry, fish, eggs dry beans and peas as alternates	2 or more serving	Iron B-vitamins
4 or more serving, including a dark green or deep yellow vegetable at least every other day.		

Contd...

The 4 food groups	*Daily amounts*	*Main nutritive contributions*
Vegetable-fruit group fruits or vegetable rich in ascorbic acid- daily Other fruits and vegetables including potatoes	A citrus fruit or other Minerals Cellulose	Vitamins
	Thiamine Niacin	
Bread-cereals group (Whole grain) ebrichesd restored)	4 or more servings	Riboflavin Iron Carbohydrates Cellulose

The 4-food group plan.

The percentage of the Daily Recommended Allowances provided by the foundation pattern will vary with individual needs. The person with a greater requirement will have those needs less well satisfied, of course, than the person with a lesser daily need. A day's dietary planned with the use of the 4-food group is illustrated in table. This menu contains more than the minimum number of servings in three of the four groups of plan. The nutrients contributed by foods in the four foundations groups in this dietary are shown in Table. Butter, jelly, mayonnaise, lemon pie, and sugar are not included in the calculations.

A comparison of the percentage of the Daily Recommended Allowances provided by foods in the 4-group plan is made. Young men obtained a lower percentage of their allowance from the foundation groups simply because the recommendations for men are higher for some nutritents than for women. However, the young man, in satisfying his energy need and promoted by a natural urge to eat, selects more of all food groups. The result is that in actual practice the diets of young men are, as a rule, more nearly adequate than those of young women.

The 4-group plan is useful as an education tool to show the foundation for an adequate diet. Because it is such a simple plan, some may not realise its importance. For the most efficient use of the 4-group plan, one should know the main nutrients contributed by the various foods in each group.

The Nutritive Value of Food Commodities: Each food group, and each food within the group, makes a particular contribution to the nutrient content of the diet. Comparison of the Nutrients Sulphide by the 4-Food Groups from the Above Menu to the NRC Recommended Daily Allowances for Young Women and Men:

Young women (18-35 years)

Young men (18-35 Years)

Some foods supply many more of the nutrients and in much greater concentration then do others. The nutritional contribution of the major food commodities, milk and milk products; meat, fish, and poultry vegetables and fruits; and cereal and cereal products, will be discussed in the following paragraphs.

Menu with the Four-Group Plan as a Guide

Breakfast		***Lunch***		***Dinner***	
Orange juice	4 oz	Swedish meat balls	3 oz	Veal cutlet	3 oz
Egg. soft-cooked	1	Green beans	½ c	Baked potato	1 med
w.w. toast	2 sl	Waldorf salad	2/3 c	Brocciol	2/3 c
Butter	1 pat	Pan roll	1	Hard roll	1
Jelly	1 tbsp	Cup custard	6 oz	Butter	1 pat
Milk, whole	8 oz	Iced tea	8 oz	Lemon pie	1 sl
		Snack -1 milk shake	Iced	tea	8 oz

Milk and its Products: Milk is considered as nature's "most nearly perfect food." Most of the known essential nutrients are found in milk, but it is a much better source of some than others. Milk and milk products are excellent sources of calcium, protein and riboflavin. Vitamin A, phosphorus, and thiamine are also supplied in good amounts. Milk and milk products are poor sources of iron and of ascorbic acid. The importance of including milk or milk products in the diet so as to obtain sufficient calcium is well known. For students, or anyone for that matter, drinking milk is one of best ways of assuming an adequate calcium intake. Each gram of milk contains about 1 mg of calcium, with a glass of milk supplying about one-third of the recommended calcium intake for the college student.

Milk contains protein of high biological value; and so complements the incomplete proteins of cereal products, fruits, and vegetables. Since milk can very well be served as a beverage at each meal of the day, a source of complete protein can therefore be assured. The protein value of milk is that each ounce contains about 1 gm of protein. Evaporated milk contains the same nutrients as whole milk. The process of evaporation, and pasteurisation also, reduces the thiamine and ascorbic acid content by about one-fifth to one-fourth. Evaporated milk is convenient to use and economical. The fortification of milk with vitamin D is a god public health measure. By adding vitamin D to milk the essential nutrients for bone formation, calcium, phosphorus, protein, and vitamin D, are brought together in one food.

The popularity of non-fat milk solids is increasing because of the lower cost, ease of storage, and lower Calorie intake. With the removal of cream from the milk, which is done prior to drying, the caloric value is reduced by about one-half. In addition, vitamin A and other fat-soluble vitamins are removed. If nonfat milk solids are used to replace the whole milk, care must be taken to provide for additional vitamin A value from other sources such as green and yellow vegetables and liver because milk is depended on to provide some of the vitamin A need. An ounce of Cheddar-type cheese provides about the same nutritive value as a glass of milk. In the manufacture of cheese some of the whey is removed, taking with it some of the water-soluble nutrients including lactose, the water-soluble vitamins (B-group), and minerals. Cottage cheese made with rennet has a higher calcium content than that made entirely by the souring of the milk (acid coagulation). In this process calcium, which is soluble in an acid medium, is carried into the whey.

Food	Measure	Energy, Cal.	Protein, gm.	Calcium, mg.	Iron mg.	Vit. A i.u.	Thiamin, mg.	Riboflavin, mg.	Ascorbic Acid mg.
Milk group									
Milk (beverage, custard, milk shake)	2½ C	400	22.5	720	0.3	875	0.2	1.05	5
Ice cream (shake)	1/8qt	145	3	87	0.1	370	0.03	0.13	1
	Total	*545*	*25.5*	*807*	*0.4*	*1245*	*0.23*	*1.18*	*6*
Meat group									
Egg, soft, ck,	1	80	6	27	1.1	590	0.05	0.15	—
Egg (custard)	1	80	6	27	1.1	590	0.05	0.15	—
Meat balls	3 oz	245	21	9	2.7	30	0.07	0.18	—
Veal cutlet	3 oz	185	23	9	2.7	—	0.06	0.21	—
	Total	*590*	*56*	*72*	*7.6*	*1210*	*0.23*	*0.69*	—

Contd...

Food	Energy, Measure	Protein, Cal.	Calcium, gm.	Iron mg.	Vit. A mg.	Thiamin, i.u.	Riboflavin, mg.	Ascorbic mg.	Acid mg.
Vegetable-fruit group									
Orange juice	4 oz	50	.5	13	0.3	245	0.11	0.03	64
Green beans	1/2c	15	1.0	31	0.4	340	0.04	0.06	8
Apples	1/2c	35	tr	4	0.2	25	0.02	0.01	1
Celery	1/4c	4	tr	10	0.1	60	0.01	0.01	2
Potato, baked	1 med	90	3	9	0.7	tr	0.10	0.04	20
Broccoli	1/2c	20	2.5	66	0.6	1875	0.07	0.15	68
	Total	*219*	*7.0*	*124*	*2.1*	*2435*	*0.32*	*0.28*	*150*
Bread-cereal group									
Bread,w.w.	*2sl*	*110*	*4*	*44*	*1.0*	*tr*	*0.10*	*0.06*	*tr*
Roll,pan	*1*	*115*	*3*	*28*	*0.7*	*tr*	*0.11*	*0.07*	*tr*
Roll, hard	*1*	*160*	*5*	*24*	*0.4*	*tr*	*0.03*	*0.05*	*tr*
	Total	*385*	*12*	*96*	*2.1*	*tr*	*0.24*	*0.18*	*tr*
Grad total		*1734*	*100.5*	*1108*	*12.4*	*5000*	*1.05*	*2.35*	*169*

Table Nutritive Value of Milk and Some Milk Products

Milk and Milk Products	*Energy Cal.*	*Protein, gm.*	*Calcium, mg.*	*Irons mg.*	*Vit. A. IU*	*Thiamine, mg.*	*Riboflavin mg.*	*Ascorbic Acid, mg.*
Milk, whole (I c)	160	9.0	288	0.1	350	0.08	0.42	2
Buttermilk, cultured, skim (I c)	90	9.0	298	0.1	10	0.09	0.44	2
Evaporated, water added (I c)	173	9.0	318	0.2	410	0.05	0.42	1
Cheese, process (I oz)	105	7.0	219	0.3	350	trace	0.12	0
Cheese, cottage, creamed, skim milk (1/2 c)	120	15.5	106	0.3	190	0.03	0.28	0
Ice cream (1/8 qt)	145	3.0	87	0.1	370	0.03	0.13	1

Table Nutritive Value of Some Meat, Fish, Poultry, and Egg

Food	*Energy Cal.*	*Protein, gm.*	*Calcium, mg.*	*Irons mg.*	*Vit. A. IU*	*Thiamine, mg.*	*Riboflavin mg.*	*Ascorbic Acid, mg.*
Beef, round (3 oz)	165	25.0	11	3.2	10	0.06	0.19	0
Beef, liver, fried (3 oz)	195	22.5	9	7.5	45,200	0.22	3.55	23
Pork, lean and fat, loin roast (2 oz)	310	21.0	9	2.7	—	0.78	0.22	0
Chicken (3 oz)	170	18.0	18	1.3	200	0.03	0.11	3
Salmon (3 oz)	120	17.0	—	0.7	60	0.03	0.16	0
Egg (1)	80	6.0	27	1.1	590	0.05	0.15	0

Value depends on amount of bone consumed.

Value: They provide protein and iron, with some phosphorus, thiamine, riboflavin, and niacin.

Meat, Fish, Poultry, and Eggs: Meat, fish and poultry are very similar in their nutritive. The protein in this group of foods, as in milk, is of high biological value. A 3-oz serving of meat supplies approximately 20 gm of protein, which is about one-third the Recommended Allowances for young adult women and over one-fourth for young adult men.

Iron is present in relatively high concentration in the organ meats, particularly liver and kidney, and in to a lesser amounts in muscle tissues. Meat is poor source of calcium . Some fish products, mainly shell fish and canned salmon, in which a portion of the bone is edible, are somewhat better than meat as a source of calcium.

Pork muscle is high in thiamine content whether other muscle meats contain only moderate amounts. The glandular organs, liver, kidney, heart, and tongue are higher than muscle meats in riboflavin content. Liver, the storage place for vitamin A, is only edible animal tissue containing any appreciable amount of this vitamin.

Eggs provide protein, iron, phosphorus, vitamin A, and riboflavin to the diet. The protein in egg, which is located both in the yolk and the white, is of good biological value. One egg contains about 6 gm of protein, approximately two-thirds the amount in one glass of milk. Phosphorus is distributed in both the white and the yolk. Iron and vitamin A are contained only in the yolk. Table gives the nutritive value of some kinds of meat, fish, poultry, and eggs.

Vegetables and Fruits: As a group, vegetables and fruits are the source of minerals, vitamins, and cellulose to the diet. Not to be disregarded is the interest they add because of their variety in colour, flavor, and texture. The great number of fruits and vegetables available for use differ markedly in nutrient content. For the sake of clarity they are grouped according to their contribution of nutrients to the diet.

Mostly the green and yellow vegetables and fruits are

outstanding in their vitamin A values. One serving of sweet potatoes, carrots, kale or cantaloup furnishes sufficient vitamin A to meet the Recommended Allowance for vitamin A for persons of all ages. In addition leafy-green vegetables provide more calcium, iron , ascorbic acid, and the B-vitamins than other vegetables.

Some of the green vegetables, spinach, chard, sorrel, and beet greens contain oxalic acid, which combines with calcium during digestion forming an insoluble salt that cannot be absorbed, and makes the calcium in those vegetables of no use to the body. However, these green vegetables, even though not supplying calcium, are invaluable in the diet for the generous amounts or iron, ascorbic acid, and vitamin A value they provide. The ascorbic acid content of broccoli, Brussels sprouts, kale, and similar vegetables is quite high.

Head lettuce, celery, cabbage, and other light green vegetables are important source for cellulose. Raw cabbage is a good source of ascorbic acid. The light green vegetables are low in vitamin A value.

Citrus fruits and tomatoes are good sources of ascorbic acid. Tomato juice contains about one-third as much as the citrus fruits. Since they are available the year round and are well liked, it is customary to include citrus fruits in most dietary plans. Other foods like strawberries, cantaloups, kale, and green peppers are also rich in ascorbic acid as citrus fruits, but due to cost, unavailability, or lack of acceptance, do not fit into the diet as a routine item as well as citrus fruits. Ascorbic acid is easily destroyed during storage, freezing, and cooking. Citrus retain their ascorbic acid content better than other foods due to the preserving action of the acids they contain.

Vegetables of the seed, root and tuber classes, like Lima beans, corn green peas, sweet potatoes, and white potatoes are higher in carbohydrate content than other varieties. Starch is stored in these areas of the plant's structure.

Vegetables and fruits as a rule provide lesser protein, fat, and total Calories. Table shows the contribution of some fruits and vegetables to the diet.

Pulses and Grains: Cereals are seeds of the grass family—wheat, corn, rice, oats, rye, and barley. From these cereal grains flours (whole wheat, white rye, barley, etc.) are manufactured for the preparation of the many varieties of breads, cakes, pastries, breakfasts cereals, macaroni products, hominy corn sirup, corn meal, and other products. Whole wheat has the following composition in terms of grams of the constituent per 100 gm of whole wheat: 72 carbohydrate, 2 fiber, 12 protein, 2 fat, 2 ash, and 10 moisture. In many countries cereals are the main food of the diet, furnishing 50 to 90 per cent of the total calories; in the United States about one-third of calories is from cereal or cereal products. More wheat and corn is grown in U.S.A. than other cereals; most of the wheat and corn is used for human consumption, a large part of the corn fed to livestock.

The cereal grains are easy to grow, easy to store, and have good keeping qualities. Whole grain cereals contribute calories (mainly in the form of carbohydrate), thiamine, iron, riboflavin, and partially in complete protein. Cereals are easy to digest, have a blend flavor, and supply roughage or cellulose.

In the preparation of cereal products for the market, parts of the cereal grain are removed. In the case of brown rice only the outer husk is removed whereas in the preparation of white rice all of the outer coats are wheat flour is made from the entire grain except for the husk; white flour is just the endosperm portion. Most of the cellulose, minerals and B vitamins are in the outer layers or coats of the grain. The endosperm portion contains the carbohydrate and incomplete protein. The germ contains most of the fat of the grain and some of the thiamine. Because the fat tends to become rancid during storage and also seems to hold a great attraction for insects, the germ is usually removed in milling. Furthermore, the public prefers the refined product for houses.

Nutritive Value of Some Fruits and Vegetables

Food	*Energy Cal.*	*Protein, gm.*	*Calcium, mg.*	*Irons mg.*	*Vit. A. IU*	*Thiamine, mg.*	*Riboflavin mg.*	*Ascorbic Acid, mg.*
Apple (1 med)	70	tr	8	0.4	50	0.04	0.02	3
Banana (1 med)	85	1.0	8	0.7	190	0.05	0.06	10
Broccoli, ck (1/2 c)	20	2.5	66	0.6	1875	0.07	0.15	68
Cabbage, ck (1/2 c)	18	1.0	38	0.3	110	0.04	0.04	28
Carrots, ck (1/2 c)	23	0.5	24	0.5	7610	0.04	0.04	5
Green beans, ck (1/2 c)	15	1.0	31	0.4	340	0.04	0.06	8
Kale, ck (1/2 c)	15	2.0	74	0.7	4070	—	—	34
Lettuce, Iceberg, (50 gm)	7	0.4	10	0.2	167	0.03	0.03	3
Lima beans, ck (½ c)	90	6.0	38	2.0	225	0.15	0.08	14
Orange juice, fresh (1/2 c)	50	0.5	13	0.3	245	0.11	0.03	64
Peaches, cn (1/2 c)	100	0.5	5	0.4	550	0.01	0.03	4
Potato, white, baked (1 med)	90	3.0	9	0.7	tr	0.10	0.04	20
Tomato juice (1/2 c)	23	1.0	9	1.1	970	0.07	0.04	20
Oatmeal, çk (1 c)	130	5.0	21	1.4	0	0.19	.0.05	0.3 0
Rice, white, enriched ck (1 c)	185	3.0	17	1.5	0	0.19	0.12	1.6 0
Macaroni, unenriched ck (1 c)	155	5.0	11	0.6	0	0.02	0.02	0.4 0
Macaroni, enriched ck (1 c)	155	5.0	11	1.3	0	0.19	0.11	1.5 0
Cake, plain (3 x 2 x ½)	200	2.0	35	0.2	90	0.01	0.05	0.1 0
Bread, white, unenriched (1 sl)	60	2.0	16	0.2	0	0.2	0.02	0.3 0
Bread, whole wheat (1 sl)	55	2.0	23	0.5	0	0.06	0.03	0.7 0
Muffin, enriched (1)	140	4.0	46	0.8	50	0.08	0.11	0.7 0
Biscuit, enriched (2 1/2" diam)	140	3.0	46	0.6	0	0.08	0.08	0.7 0
Pan roll, enriched (12 per Ib)	115	3.0	28	0.7	0	0.11	0.07	0.8 0

The protein in cereal products varies in biological efficiency but as a whole is classified as partially incomplete. However, with supplementation or protein from milk or meat or eggs, the cereal proteins make a significant contribution to the diet. About 20 per cent of the protein in diet is from cereal sources; on low-cost dietaries the percentage is relatively higher. Milk and cereal make a good nutritional team, cereal supplying iron, thiamine, and partially complete protein, and milk providing calcium, vitamin A, and complete protein. The use of the milk solids (dried milk) in bread improves the nutritive contributions of this product. It is important that whole grain, restored or enriched cereals, and cereal products be selected for the diet in order to benefit from their greater nutritive contributions. Table gives the nutritive contributions of some cereal products to the diet.

Energetic Oxidisation: Energy for physiological activity is ultimately obtained from the sun. The sun gives out energy in the form of visible light waves and invisible ultravoilet rays. When these impinge on green plants, the latter are able to absorb the energy and build-up compounds and such as sugar and starch from the carbon dioxide of the air and the water of soil. The compounds become storehouses of energy within the green plants.

By the action of oxygen on these carbon compounds, they can again be broken down through a series of stages into carbon dioxide and water and, in the process, the stored energy is once more liberated. This breakdown happens in the plants themselves when they respire and is the source of energy used by them for growth, development and reproduction. But far more energy is stored in starch and sugar than is used by the plants and, when animals eat the plant products, this energy becomes available to them.

Animals, too obtain oxygen from the air in breathing. This is transferred in the lungs to the blood and is carried to all parts of the body. Food which is eaten, digested and absorbed is also taken in the circulation to organs where it can be temporarily stored. Most of the starch and sugar eaten is stored in the form of another carbohydrate, glycogen, in either the liver or the muscles, and as we have seen, is broken down to provide energy. The recognition of these facts led to a great deal of work on the quantities of food needed to produce the energy for certain tasks and it was found that the energy value of a food could be expressed in terms of a heat-unit or calorie. A calorie is the amount of heat-energy required to raise the temperature of 1 g. of water through 1 °C. This was found too small a unit to be convenient for expressing the energy values of foods, so a unit was chosen of 1,000 calories or 1 large Calorie (spelt with a capital C). The next step was to find out how many calories were required by individuals of different weights, both resting and carrying out tasks with different energy requirements.

The following are average values:

	Calories
Metabolism of sleep (8 hours)	518
Basal metabolism while awake (16 hours)	1,152
Allowance for light work	1,200
Total	2,870

(Best and Taylor, 1950).

Daily Calorin Allowances for Adults

Requirment	*Description*	*Calorie requirement*
0	No work, almost basal (*e.g.* lying in bed).	1,750
1	Sedentary work (30 Cals/hr.) and little travelling (65 Cals.)	2,250
2	Light work (70 Cals. /hr.) and travelling (130 Cals.)	2,750
3	Medium work (100 Cals./hr.) and travelling (130 Cals.	3,000
4	Heavy work (200Cals. / hr.) and travelling	(130 Cals.)
	Extremely heavy work (450 Cals./hr.) and travelling (130 Cals.)	5,000
0	No work, almost basal (e.g lying in bed)	1,500
1	Sendentay work (30 Cals./hr.) and little travelling (50 Cals.)	2,000
2	Light work (70 Cals./hr.) and travelling (100 cals.)	2,250
3	Medium work (100 Cals./hr.) and travelling (100 Cals.).	2,500
4	Heavy work (200 Cals./hr.))	
5	Very heavy work (200 Cals./hr.) and travelling (100 Cals.)	3,750

The recognition that the quantity of food required could be exactly measured so captured the imagination that for many years no thought was given to the quality of the food supplied. Calories were all important and the neat way in which a man's energy needs could be expressed numerically no doubt delayed the recognition of the more subtle needs for particular factors in foods.

This is doubtless one reason why vitamins were not discovered until the present century, and why are still ignorant about certain aspects of nutritional requirements. It is, however, established that the basal needs for calories, are best met by a balance between proteins, fats and carbohydrates:

other needs are adequate amounts of fluid, certain mineral salts and vitamins. Although this book is mainly concerned with vitamins, it is not possible to study them without reference to other nutrients.

The Table shows present-day views of human requirements of vitamins and calories. The figures given in the table are, of course, averages. As with all physiological functions, there are great individual variations but these can be discovered only by experience. However, an example will serve to illustrate the point.

Mr. A, a lean, bony man, has a hearty appetite and seems able to eat as much as he likes without showing any tendency to grow fat. By contrast Mr. B, who is also physically active, puts on weight without apparently eating to excess. These two seem to have different basal mechanisms for the use of food, one making his calories go further, as it were, than the other.

The same seems to be true, to some extent for vitamin needs. Among people in a group living on a diet deficient in one or other of the vitamins, some will show signs of disease, others, whose needs are evidently less, will escape. Both nicotinic acid and nicitinatmide are physiologically active as what was originally the P-P factor.

To·avoid confusion with nicotine (which has no vitamin activity) the term Niacin was introduced in the U.S.A. and has been used in the table compiled by the B.M.A. Committee on Nutrition.

In planning diets, however, the table given can safely be used as a guide. Any individual idiosyncrasies in the nature of special needs must be allowed for separately.

Methods of Assessing the Adequacy of Diets: How can you assess that a diet which has been planned or eaten is adequate in nutrient content?

There are various procedures for assessing the nutritive value of the diet. The most simple of these is the use of a "score card."

The use of a food pattern as a guide in planning an adequate diet has already been discussed, and score cards have been developed for determining how well diet meets the suggested pattern. A more precise method of evaluating the nutritive adequacy of diets is computing the nutrient content of each food using a food composition table.

The most precise method of all is actual laboratory determination of the energy, protein, mineral, and vitamin content.

The choice of the procedure for evaluating the energy and nutrient content of a diet depends on the use to be made of the findings and the accuracy with which the food consumption data were collected.

Vitamins in Nutrition and Health

Report of the Committee on Nutrition- B.M.A. 1950. p. 22 and p. 23 (extract)

Age and sex	Requirement class	Calories	Vit. A &- carotene i.u. daily	Vit. D i.u. daily	Thia min mg. daily	Niacint mg. daily	Ribo flavin mg. daily	Vit. C mg. daily
Both								
0-1		1,000	3,000	800	0.4	4	0.6	10
2-6		1,500	3,000	400	0.6	6	0.9	15
7-10		2,000	3,000	400	0.8	8	1.2	20
Males								
11-14		2,750	3,000	400	1.1	11	1.6	30
15-19		3,500	5,000	400	1.4	14	2.1	30
sedentary work		2,250	5,000		0.9	9	1.4	20
	medium work	3,000	5,000		1.2	12	1.8	20
	heavy work	4,250	5,000		1.7	17	2.6	20
Females								
11-14		2,750	3,000	400	1.1	11	1.6	30
15-19		2,500	5,000	400	1.0	10	1.5	30
	sedentary work	2,000	5,000		0.8	8	1.2	20
	medium work	2,500	5,000		1.0	10	1.5	20
	very heavy work	3,750	5,000		1.5	15	2.2	20
Pregnancy								
	1st half	2,500	6,000	400	1.0	10	1.5	40
	2nd half	2,750	6,000	600	1.1	11	1.6	40
	Lactation	3,000	8,000	800	1.4	14	2.1	50

Vitamin B_1, aneurin and thiamine are synonymous but the most recent recommendation is that Thiamine should be adopted as the chemical name of vitamin B_1.

Use of Dietary Score Cards: Dietary score cards set up for evaluating the diet on the basis of food groups are used mainly as a basis for nutrition education programs (Tables). A limitation of each score card is its failure to give the complete nutritive contribution of each food.

On the other hand, it is a commendable educational tool in that it encourages a varied diet and provides a good "rule of thumb" for the selection of an adequate diet.

Use of Food Composition Tables: By far the most common method of determining the nutrient content of the diet is the use of food composition tables. In various chapters of the book of the food tables have been used in the study of the energy value and content of the different nutrients of various foods. The values are obtained by laboratory analysis with each representing many determinations.

The composition of the same food varies widely, specially in fruits and vegetables. Growing conditions, variety, fertilizer treatment, care in handling, all make a difference. Preparation of the food for the table preservation, and storage also have an influence on the composition of most foods.

A Dietary Score Card

Foods	*Each Day You Need*	*Score for Your One Day Score*
Green and yellow vegetables (raw or cooked in small amount of water)	1 serving	10
Oranges, grapefruit, tomatoes, or other vitamin-C-rich food	1 serving	10
Potatoes and other vegetables and fruits	3 servings	5
Milk and milk products	Children, 3-4 cups Adolescents, 3-4 cups Adults, 2 cups	20

Contd...

Foods	***Each Day You Need***	***Score for Your One Day Score***
Meat, poultry, or fish	1 serving	15
Meat, poultry, fish, or meat alternates (dried beans, peas, peanut butter)	1 serving	10
Eggs	1 daily (at least 4 a week)	5
Cereal, whole grain or enriched	1 serving or 2 slices bread	
Bread, whole grain or enriched	1 or 2 slices at every meal	5
Butter or other fats	2 to 3 level tbsp butter or enriched margarine	5
A good breakfast, including some form of protein, as milk or egg		10
	Total	100

Those who have compiled food tables have had to decide on the values that they considered most representative for inclusion in the table. The necessity of exercising judgment in preparing a food table accounts for some of the differences that exist among the compilations of food composition values. As a rule the most food table includes new data add to the accuracy.

Short Methods Using a Food Composition Table: Short method for dietary calculation have been developed in which foods similar in composition are grouped, together, a representative value for each of the nutrients is calculated, and the values listed in a table. Totalling the amount of foods of like composition in a diet and computing the nutritive contribution of the group is shorter, of course, than computing each food separately. That is the manner in which the short methods save time in calculation.

The short methods are found to agree with the long or

usual method if applied to a varied diet, and if the food habits do not differ widely from the usual pattern of the dietary habits of the people of a country as a whole.

In regions where there is a relatively higher intake of certain foods, such as corn meal in the southern region, a revision in the cereal and bread portion of the table becomes necessary to obtain an accuracy approaching that of calculating each food individually. For survey studies the shorter method of dietary calculation has been found satisfactory. Time required for computation can be reduced without great sacrifice to accuracy.

Chemical Analysis: In research studies designed to determine quantitatively the nutrient intake as part of a metabolic balance study, the composition of a food is determined by laboratory analysis rather than by use of a food composition table Laboratory Analysis is the most accurate of any procedure for determining the nutritive value of food, since the nutrient content of some foods varies widely.

To illustrate, the ascorbic acid content the thirteen different varieties of potatoes ranged from 8.2 to 17.4 mg. per 100 gm. of fresh weight, and for thirty-five varieties of cabbage from 32.4 to 100.7 mg. per 100 gm. of fresh weight. It should be recognised, however, that the fruits and vegetables vary more in composition than most other foods. The differences between computed and analysed values for given diets have been studied. It is generally concluded that agreement is sufficiently close to warrant the use of food tables in dietary surveys, reserving laboratory analyses for controlled studies.

Budget of Family

Ultimately, daily menus, a market list, and a plan for purchasing food must be made by families who maintain households and by agencies who make allowances for families in need of assistance. The Consumer and Food Economics

Research Division of the U.S. Department of Agriculture prepares materials that are of genuine assistance as guides in planning food budgets.

The quantities and kinds of foods needed to adequately nourish different individuals are estimated at low-cost, moderate-cost, and more expensive diet. The Research Division has prepared plans for twenty age-sex groups and for eleven food commodity groups. These groups include milk and milk products; meat, poultry, and fish, eggs, dry beans, peas, and nuts; grain products; potatoes; citrus fruits and tomatoes; dark green and dark yellow vegetables; other vegetables and fruits; fats and oils; sugars and sweets.

To plan a food budget for a family of a certain size it is necessary to add together the estimated quantities needed for each family member. For example, a week's food plan for a family of a man and a woman both 33 years of age, a girl 8, and a boy 11 would be as summarised in Table.

The student of nutrition is aware that merely planning for kinds and amounts of food is not enough; the planning is made best with knowledge of the food preferences of the family and the availability and price of these foods. The budget plan allows for choice of foods within the commodity groups. Further adjustment can be made even between the groups with a knowledge of food values. For instance, potatoes could not substitute for eggs nutritionally, but another serving of meat would substitute very well.

The estimated cost of one week's food supply for a four-member family (a girl 8, a boy 11, man and woman each 33 years of age) as of January 1963 was $32.80 for the moderate-cost plan.

It is gratifying that an adequate diet can be provided at different cost levels. The low-cost plan contains substantially more potatoes and grain products. The lower-cost meat cuts can be used in the limited-expense plan with the assurance that the nutritive value is equal to that of the higher-priced cuts. The moderate-cost and liberal plans include more milk, eggs, meat, fruits and vegetables. More expensive items such as foods out of season and the more highly processed foods are allowed in the moderate-cost and liberal plans.

Quantity of Food for a Family of Four for One Week

Food Groups	*Unit*	*Low-Cost Plan*	*Moderate Cost-Plan*	*Liberal Plan*
Milk, milk products	qt	16.5	17.5	19.0
Meal, poultry, fish	lb	11.5	17.2	20.8
Eggs	no.	25.0	29.0	29.0
Dry beans, peas, nuts	lb	1.4	.9	0.9
Grain products	lb	12.5	11.5	10.9
Potatoes	lb	10.0	8.5	7.8
Citrus fruit, tomatoes	lb	7.5	9.0	11.5
Dark green, deep yellow vegetables	lb	3.5	3.5	3.5
Other vegetables, fruits	lb	19.8	22.5	25.2
Fats, oils	lb	3.1	2.8	2.8
Sugars, sweets	lb	3.0	3.9	4.6

Cost of one Week's Food at Home[1] Estimated for Foods Plans at Three Cost Levels, June 1964, U.S.A. Average

Sex-Age Groups[2]	*Low-Cost Plan, Dollars*	*Moderate Cost Plan, Dollars*	*Liberal Plan, Dollars*
Families			
Family of two 20-35 years[3]	14.60	19.60	22.70
Family of two, 55-75 years[3]	12.20	16.50	18.70
Family of four, preschool children[4]	21.40	28.40	32.80
Family of four, school children[5]	24.60	33.00	38.30
Individuals[6]			
Children under 1 year	2.90	3.80	4.10
1-3 years	3.70	4.80	5.50

Contd...

Sex-Age Groups[2]	*Low-Cost Plan, Dollars*	*Moderate Cost Plan, Dollars*	*Liberal Plan, Dollars*
3-6 years	4.40	5.80	6.70
6-9 years	5.20	7.00	8.30
Girls, 9-12 years	6.00	8.00	8.90
12-15 years	6.60	8.80	10.20
15-20 years	6.90	9.00	10.20
Boys 9-12 years	6.10	8.20	9.40
12-15 years	7.10	9.70	11.00
15-20 years	8.30	10.90	12.60
Women, 20-35 years	6.20	8.30	9.40
35-55 years	6.00	7.90	9.10
55-75 years	4.70	6.90	7.80
75 years and over	4.70	6.20	7.20
Pregnant	7.50	9.60	10.80
Nursing	8.60	11.10	12.30
Men 20-35 years	7.00	9.50	11.20
35-55 years	6.60	8.80	10.20
55-75 years	6.00	8.10	9.20
75 years and over	5.60	7.80	8.90

Food Values and Food Costs: The buying of food is the largest single expense in the budget for most families. An average of 19 per cent capita of disposable personal income is spent for food in the United States. With increasing income, the per cent going food diminishes.

Economical purchase of food makes it possible to provide adequate diets at lower cost. Some nutrients are provided at less expense by one food or group of foods than by others. The approximate percentage of the food dollar spent for milk and milk products; meat, fish, poultry, eggs, dry beans, and nuts; vegetables; fruits; and grain products is shown in Fig. Other foods not included in Fig. are fats, sugar, and sweets, and miscellaneous items, which account for about 20 per cent of the total food expenditure. The protein group of foods is the most expensial, amounting to 37 cents of the food dollar or 37 per cent of the food expenditure. Milk and milk products account for 15 cents of the food dollar, vegetables 12 cents, grain products 11 cents, and fruits 8 cents.

Foods differ in the economy with which they provide the nutrients of the diet. Protein, calcium, vitamin A value, and thiamine were selected to illustrate this point. Using Fig. as a guide, in each food group if the nutrient bar exceeds the cost bar in length, that food group is a good buy for providing that nutrient, because the per cent of the expenditure for that group of foods is less than per cent of the nutrient obtained. The relative lengths of the bars will vary, of course, with price. It can be clearly seen that the milk group of foods is an excellent buy for calcium but only fair for supplying thiamine; the meat, fish, poultry, eggs, dry beans, and nuts group is a good buy for protein but a poor buy for calcium; vegetables are excellent for vitamin A, poor for protein; fruits are not a good buy for the nutrients listed, but had ascorbic acid been selected as one of the nutrients in the illustration, fruits would appear as good buy for that nutrient.

Grain products are a good buy for protein and thiamine, and it may be surprising to note, for calcium, accounted for in part by the use of bread containing non-fat milk solids. Of course, little if any vitamin A value is furnished by grain products. The protein of cereals is economical to purchase and becomes of the good nutritional quality when supplemented with milk or other sources of complete protein. Grain products are a good buy for protein because, as shown in the cost bar is shorter than the protein bar.

3

Elements of Food

Treatment by diet or the Diet therapy is concerned with the modification of the normal diet to meet the requirements of the sick individual. Its purposes are:

- To maintain good nutritional status.
- To correct deficiencies.
- To afford rest to the whole body or the body that is affected.
- To maintain body's ability to metabolise the nutrients.
- To bring about changes in body weight whenever necessary.

Diet therapy in most instances is not a remedy in itself but is a measure which supplements or makes the medical or surgical treatment more effective.

Therapeutic nutrition begins with the normal diet. Advantages of using normal diet as the basis for therapeutic diets are:

- It emphasises the similarity of psychologic and social needs of those who are ill and those who are well, even though there is quantitative and qualitative differences in requirements.

- Food preparation is simplified when the modified diet is based upon the family pattern and the number of items required in special preparation is reduced to a minimum.
- The calculated values for the basic plan are useful in finding out the effects of addition or omission of certain foods, for example, if vegetables are restricted vitamin A and C deficiency can occur.

Compulsory Ingredients

The alteration of the normal diet requires an appreciation of (1) The underlying disease conditions which require a change in the diet, (2) The possible duration of the disease, (3) the factors in the dietary which must be altered to overcome these conditions, and (4) the patient's tolerance for food by mouth. In planning meals for a patient his economic status, his food preferences and his occupation and time of meals should also be considered.

The normal diet may be modified (1) to provide change in consistency as in fluid and soft diets: (2.) to increase or decrease the energy value: (3) to include greater or lesser amounts of one or more nutrients, for example, high protein, low sodium, etc.; (4) to increase or decrease bulk-high and low fibre diets; (5) to provide foods bland in flavour.

The planning of a therapeutic diet implies the ability to adapt the principles of normal nutrition to the various regimens for adequacy, correctness, economy, and palatability. It requires a recognition of the need for dietary supplements such as vitamin and mineral concentrates when the nature of the diet itself imposes severe restrictions, the patient's appetite is poor, or absorption and utilization are impaired so that the diet cannot meet the needs of optimum nutrition.

Dietary history should serve as the basis for planning each diet. The dietary history will reveal the patient's past habits of eating with respect to dietary adequacy, likes and dislikes, meal hours, where meals are eaten, budgetary problems, ability to obtain and prepare foods.

The likes and dislikes of patients are respected because food habits are deep-seated and it is not possible to change them overnight. It requires considerable encouragement and understanding on the part of the doctor-nurse-dietitian team to bring about important changes in the diet. Intelligent planning of therapeutic diets necessitates a consideration of food costs, the avoidance of waste, and retention of nutrient values so that the diet is economically practicable.

Nutrition in Nursing Homes

Clear-fluid Diet: Whenever an acute illness or surgery produces a marked intolerance for food as may be evident by nausea, vomiting, anorexia, distention and diarrhoea, it is advisable to restrict the intake of food. In acute infections before diagnosis, in acute inflammatory conditions of the intestinal tract, following operations upon the colon or rectum when it is desirable to prevent evacuation from the bowel, clear fluid is suggested. This diet is also given to relieve thirst, to supply the tissues with water, to add in the removal of gas.

The diet is made up of clear liquids that leave no residue, and it is non-gas forming, non-irritating and non-stimulating to peristalitic action. This diet is entirely inadequate from a nutritional standpoint since it is deficient in protein, minerals, vitamins, and calories. It should not be continued for more than 24 to 48 hours.

The amount of fluid is usually restricted to 30 to 60 ml per hour at first, with gradually increasing amounts being given as the patient's tolerance improves. This diet gives 300 k.cal. and no protein.

This diet can meet the requirement of fluids and some minerals and can be given in 1 to 2 hour intervals.

Full-fluid Diet: This diet bridges the gap between the clear fluid and soft diet. It is used following operations, in acute gastritis, acute infections and in diarrhoea.

This diet is also suggested when milk is permitted and for patients not requiring special diet but too ill to eat solid or semisolid foods.

In this diet foods which are liquid or which readily become liquid on reaching the stomach are given. This diet may be made entirely adequate and may be used over an extended time without fear of deficiencies developing, provided it is carefully planned. This diet is given at 2-4 hour intervals. This diet gives 1200 k.cal. and 35g. of protein.

Soft Diet: This is one of the most frequently used routine diets; many hospital patients are placed on this until a diagnosis is made. It bridges the gap between acute illness and convalescence. It may be used in acute infections, following surgery, and for patients who are unable to chew.

The soft diet is made up of simple, easily digested food and contains no harsh fibre and no rich highly seasoned food. It is nutritionally adequate when planned on the basis of normal diet. Patients with dental problems are given mechanically soft diet.

It is often modified further for certain pathologic conditions as bland and low residue diets. In this diet, three meals with intermediate feedings should be given. This diet gives 1500 k.cal. and 35-40 g. of protein.

Regular Normal Diet: It is most frequently used in all hospitals. It is used for ambulatory and bed patients whose conditions does not necessitate a special diet of one of the routine diets. Many special diets progress ultimately to a regular diet. Table below gives contents allowed for soft diet, full fluid and clear fluid diet.

Contents of Soft, full-fluid and Clear-fluid Diets

Types of food	*Foods allowed*		
	Soft diet	*Full-fluid*	*Clear-fluid*
Cereals	Refined, finely ground whole grain	Gruels, porridges kanji, ragi malt	Barley water
Pulses	All dals	Dal soups, dal payasam	Dal water
Vegetables and Fruits	Juices, pureed, cooked and mashed, baked, ripe banana	Strained juices, cooked and pureed fruits	Clear strained fruit juice
Milk	Milk and milk products,	Milk and milk beverages, cheese, fine cream	Whey water milk shakes, lassi
Fats and oils	Butter, oil, cream, margarine	Butter, oil and cream	—
Meat and fish	All except pork, minced fish poultry	—	—
Eggs	All except fried	Only in beverages beaten in fruit juices	Egg white well
Sugar, jaggery	Sugar or glucose	jaggery	Sugar and All and glucose
Nuts and oil seeds	None	None	None
Beverages	All	Tea, coffee, egg, non-carbonated beverages	Tea, coffee (without milk) carbonated beverages, coconut water
Soups	All	Strained	Fat free broth,
Desserts	Custards, kheer, puddings	Custards, ice-cream plaon gelatin	Plain gelatin

The regular hospital diet is simple in character and preparation, ease of digestion, and calculated to afford maximum nourishment with minimum effort to the body. The diet is well balanced, adequate in nutritional value and attractively served to stimulate a possible poor appetite. This diet gives 1800-2000 k.cal. and 42-45 g. of protein.

Techniques of Diet

Enteral nutrition can be provided either orally or by tube feeding. By definition enteral means "within or by the way of the gastrointestinal tract". In practice enteral nutrition is generally considered as tube feeding.

Oral feeding is the best for the nourishment of a patient. But in the following conditions it is not possible to give the feeding orally and tube feeding or parenteral feeding is resorted.

- Those who cannot swallow due to paralysis of the muscles of swallowing (diptheria, poliomyelitis) or cancer of the oral cavity or larynx. Those who cannot be persuaded to eat.
  - Those with persistent anorexia requiring forced feeding.
  - Semiconscious or unconscious patients.
  - Severe malabsorption requiring administration of unpalatable formula.
  - Short bowel syndrome.
  - Those who are undernourished or at risk of becoming so.
  - Those who cannot digest and absorb.
  - After surgery patients with neurological and renal disorders or have continued fevers or diabetes.
- Babies of very low birth weight.

Feeding with Tube: This is done by passing a tube into the stomach or duodenum through the nose which is called nasogastric feeding or directly by surgical operation known as gastrostomy and jejunostomy feeding. A satisfactory tube feeding must be :

1. nutritionally adequate
2. well tolerated by patient so that vomiting is not induced.
3. easily digested with no unfavourable reactions such as distension, diarrhoea or constipation
4. easily prepared and
5. inexpensive.

Type of foods supplied through the tube may be:

a. natural liquid foods

b. blenderised to make liquid food

c. commercially supplied polymeric mixtures or elemental diet (predigested diet)

Juice of Fruits

Natural liquid foods like whole or skim milk, eggs and some form of carbohydrate such as strained cooked cereals, sugar or molasses can be given.

Vegetable oil or cream and non fat dry milk are also incorporated to increase the calorie and protein levels respectively.

Foods like 'Complan' can be added to different foods to increase the nutritive value. Many convalescent patients who are often depressed and have little appetite benefit from appetising foods of high nutrient density.

Electronic Devices and Diet

In 'blenderised feeding' the ordinary food items which cannot be swallowed are blended to make thin liquid which can pass through nasogastric tube.

Food mentioned in Table below is cooked before putting in an electric mixer.

- Water is added to make the volume 1500 ml
- Gives 1500 calories.
- Each ml provides one calorie.
- Gives 50 g protein.
- Cost is no more than an average meal in hospital.

Composition of Blenderised Food for Tube Feeding for Hospital Practice

Ingredient	*Amount g.*
Rice	75
Green gram dal	40
Bread	20
Milk	200
Skim milk powder	60
Spinach	50
Pumpkin	50
Carrots	50
Banana	70
Sugar	60
Refined oil	20
Butter	7

Tube feedings are prepared from the ordinary foods of a normal diet by using a high speed blender are generally preferred to other types of formulas.

Blenderised feeding can also be prepared by using starch, milk, skim milk powder, egg, soya flour, refined oil, salt and curd. Blenderised tube feedings are well tolerated and are only infrequently associated with diarrhoea.

They are less expensive than commercial formulas. (c) Normal foods may be replaced by preparation of protein or amino acids, fats and carbohydrates in easily assimilable forms. Such preparations are known as elemental diets. For tube feeding commercially supplied polymeric mixtures can also be used.

'Complan' is a commercial mixture of milk protein casein, carbohydrates (lactose) and fat. Efficiency of the kidney may not be affected if 'Complan' is given to supply 2000 kcal.

The commercial preparations possess the advantage of convenience, constant composition, presterilisation, minimal preparation time and ease of administration.

The disadvantages are high cost and unpleasant taste and sometime high osmolarity. Elemental diets though can be given orally, palatability may be a problem and tube feeding can be resorted.

Though elemental diets are better tolerated the human gut is capable of absorbing complex moieties just as well.

Dietary Conditions

A concentration of about 1 kcal. per millilitre is satisfactory. Lesser concentration increases the volume which must be given to meet the nutrient and energy needs.

Greater concentration is more likely to produce diarrhoea and may be too thick to pass through a nasogastric tube.

The feeding is started through a continuous drip at a rate of 50 ml per hour.

The rate is increased by 20 ml every 24 hours until the required volume is achieved, usually with 100-120 ml per hour. The concentration or rate of flow may have to be decreased, if there is vomiting, abdominal cramps or diarrhoea.

Feeding requirements are based on previous nutritional status and other feeds given to the patient.

Solutions of Nutrition

Feeding solutions have to be treated with full hygienic precautions during the preparation, storage and administration. Feeds should be stored in a refrigerator to avoid bacterial growth and taken out before administration in time to reach room temperature; very cold feeds are not tolerated.

A feed should be discarded when it has been more than 24 hours, out of storage.

Feeding Requirements in Tube Feeding

Nutrient	Amount
Fluids	30 ml/kg
Energy	32 kcal/kg
Protein	1g/kg body weight
Sodium	30-40 mMol (provided there are no external losses)
Potassium	1 mMol/gram of protein

Vitamins and minerals supplementation should be given.

There are four different infusion techniques:

1. Continuous method is to feed via an infusion pump or gravity flow.
2. Cyclic method has off and on periods in 24 hrs, *e.g.*, 8 hrs on and 16 hrs off.
3. Interrupted method is a regular period of interruption in 24 hrs, *e.g.*, 4 hrs on and 4 hrs off.
4. Bolus method where large volumes are given in a short time, *e.g.*, 200cc in a minimum of 10 min time.

Record Keeping

Nursing staff should accurately record (1) the time when a feed is started and completed (2) the volume administered (3) water used to irrigate the tubing and (4) the patient's output of urine. Careful monitoring is needed to see that patient is in fluid balance.

The Bottlenecks

The following checks should be carried out as a routine:

1. Initially, measurement of plasma urea, electrolytes and blood glucose.
2. Every eight hours, a record of fluid intake and output with a report if urine output falls to 30 ml/hour.

3. Daily, calculation of actual intakes of energy, protein, fat and carbohydrate.
4. Every second day, weighing should be done early in the morning. If the patient is in fluid balance and weight is decreasing, extra protein and energy should be added to the regime.
5. Weekly collection of urine for 24 hours for a rough estimation of nitrogen balance, a full blood count and measurement of mid-arm circumference and triceps skinfolds.

Advantages of Tube Feeding

Nasogastric feeding through a tube has many advantages over parenteral feeding.

- Adequate nutrition could be easily given by this method.
- Foods and drugs which may not be liked by the patients can be administered.
- Large quantities of fluids can be given daily with safety.
- The dangers like electrolyte and trace element imbalance of parenteral feeding are avoided.
- The gastric contents can be removed periodically and examined.
- Comparatively economical.
- Gives rise to less complications.
- Better tolerated and physiologically safer.
- Maintains gut integrity thereby preventing translocation of bacterial toxins from the gut into the circulation.

Parenteral Feeding: Here the nutrient preparations are given directly into a vein. This method may be used to supplement normal feeding by mouth but can provide all the nutrients necessary to meet a patient's requirements. Then it is known as total parenteral nutrition or TPN.

Total parenteral nutrition is defined as provision of all nutrients essential for normal homeostasis and growth in the required amounts through parenteral route.

The same process is called hyperalimentation when atleast 150 per cent of daily requirements are provided to produce a positive nitrogen balance for gain in weight.

Partial parenteral nutrition provides 30-50 per cent of daily nutrients while the rest is provided through enteral route. The daily requirements of a child which have to be mandatorily provided in any TPN programme are:

- glucose
- emulsified fat
- crystalline amino acids
- vitamins including B_{12} folic acid and vitamin K
- electrolytes: sodium, chlorine, phosphorus, potassium, calcium and magnesium
- trace elements: zinc, copper, chromium, manganese and iodine
- water

Total parenteral nutrition provides all nutrients in a concentrated form to avoid fluid overload. Adequate non-protein calories are provided in the programme so that induced amino acids are used only for protein synthesis (150 NPC/g of nitrogen).

TPN Formula for Children: Calories: Calorie requirements vary with age, size, physical activity, state of health, basal metabolism, growth and SDA of food.

A rule of the thumb would be 110-125 kcal./kg/day for newborn 100/110 kcal/kg/day for older child.

Hypertonic Glucose: 50%, 25%, 20% and 8% are used for providing calories. Glucose more than 15% requires central vein for infusion. Maximum rate of infusion should be 6 mg/kg/mt. Upto 70% of requirements can be met with glucose. Providing more glucose than recommended only leads to fat and water accumulation. Addition of potassium and insulin prevents loss of sugar in urine. 15-30 g/kg/day can be provided by glucose.

Isotonic Fat Solution: Provides more calories with low osmolarity as calories are provided in concentrated form. So peripheral vein can be used. It is not lost in urine.

Amino Acids: 2-3 g/kg/day should be given. One litre of Astymin gives approximately 90 g of amino acids. The solution can be mixed with routine glucose based maintenance solutions. Adequate calories as glucose and lipid have to be provided to prevent amino acids from being used up for energy production.

Electrolytes: 3.5 mEq of sodium/potassium/chloride to be provided daily.

Vitamins: Standard multivitamin preparations are available. They can be added to the drip. Vitamin K and B_{12} can be given intramuscularly. Folic acid can be added separately.

Trace Elements: Oral preparation called Aquamin which contains all the major trace elements can be given (1-3 tsp/day). Composition of standard trace elements solution, TPN requirements are given in tables below.

Standard Trace Elements Solution

Trace element	*Amount*
Zinc sulphate	800 mEq
Copper sulphate	400 mcg
Sodium fluoride	20 mcg
Sodium iodide	118 mcg
Manganese sulphate	400 mcg/litre

Summary of TPN Requirements for Children

Age	*Water ml/kg*	*Dextrose g/kg*	*Amino acids g/kg*	*Lipids g/kg*
Neonate	80-150	7-15	2-5	1-3
Infant	120-200	12-30	2.5-4	1-3
Older child	80-150	7-15	1.5-3	1-3

Iron is not to be used in the presence of infections due to potentiation of gram negative septicaemia-multiplication of bacteria in blood stream.

Guidelines for protein, carbohydrate and fat are the same for enteral and total parenteral nutrition.

TPN Formulas for Adults

TPN solution is given 100 ml per hour. For 24 hours, 2400 ml is given.

Calories from Carbohydrates

TPN solution contains 25% dextrose	
That is, 100 ml contains	25 g
2400 ml contains	600 g
1 g of dextrose in monohydrate form gives	3.4 kcal
600 g of dextrose gives	3.4 x 600
	= 2040 kcal

Calories from Protein

TPN solution contains	4% amino acids
That is 100 ml contains	4 g
2400 ml contains	4 x 24 = 96 g
1 g of amino acid gives	4 kcal
96 g of amino acid gives	4 x 96
	= 384 kcal

Calories from Fat

Requirement of fat is	50 g
10% fat emulsion	1.1 kcal
500 ml of 10% fat emulsion	1.1 x 500
	= 550 kcal
Total calories given in TPN solution is	
Calories from carbohydrate	2040
Calories from protein	384
Calories from fat	550
	2974 kcal

Total calories given in TPN solution is 2974 kcal.

Table gives adult requirements of minerals for TPN:

Table Adult Requirement of Minerals for TPN

Mineral	*Amount*
Calcium	10-15 mEq/day
Phosphorus	10-15 mMol/1000 carbohydrate calories
Magnesium	15-20 mEq/day
Zinc	2.5-6.5 mg
Copper	0.5-1.5 mg
Chromium	10-15 mcg
Selenium	30-100 mcg
Manganese	0.15-0.8 mg

Patient Monitoring while on TPN: Daily weight, weekly skin-fold thickness, mid-aram circumference, length and head circumference are monitored.

Table below gives frequency of biochemical monitoring in TPN.

Biochemical Monitoring for a Patient on TPN

Daily	*Weekly*	*Twice weekly*
Dextrostix (4 times)	Total protein	Haemotocrit
Blood glucose	Creatinine	Haemoglobin
Urine sugar (4 times)	Blood culture	Urea
and Acetone		Ammonia
Electrolytes		Lipid profile
Acid/Base balance		Calcium, phosphorus
		Candida culture

The Problems

Internal nutrition maintains intestinal epithelium, whereas parenteral nutrition causes villous atrophy.

Catheter related cardiac arrhythmias, air embolus, thrombosis, perforation and sepsis can occur. Metabolic complications like hyperglycaemia, osmotic diuresis, azotaemia (nitrogen in blood), electrolyte disturbance, acid base change, fatty degeneration of liver can also occur.

Nutritional complications like loss of lean body weight, vitamin deficiency, vitamin toxicity, increase in fat and water, essential fatty acid deficiency and trace metal deficiency can also occur.

Table below gives sample TPN regimen.

Sample TPN Regimen (glucose-lipid based)

Age: 8 months	weight: 5 kg
Requirements	
Energy	500 kcal
Protein	15g
Water	800-900 ml
Multivitamins	
Trace elements	

Nutrient	*Water ml*	*Energy kcal*	*Protein g.*
• 25% intralipid solution	150	150	
• 15% glucose paediatnic solution	600	350.	
• Astymin	150		15
Total	900	500	15

Vitamin K* (IM)	1cc (once/15 days)
Inferon (IM)	1cc/week
Trace metal solution (IV)	2 ml/daily
VITAMIN B_{12}	½cc once/month
Oral folic acid*	
*not present in standard solution	

Derick and Ruberg summarised that "TPN is given to all patients who cannot eat, should not eat, will not eat and can not eat enough."

Special Feeding Methods should be Avoided for Three Reasons

(a) Supplementary, tube and intravenous feeding are each more expensive than normal feeding. So they should not be used when the patient can get adequate nourishment from an appetising diet, well served and with coaxing to eat it.

(b) Tube and intravenous feeding inevitably cause inconvenience and some discomfort to a patient.

(c) Although tube feeding is safe, atleast in comparison with a prescription for most drugs, parenteral nutrition carries complications.

ISPEN

Indian Society for Parenteral and Enteral Nutrition (ISPEN) is an association that is involved in nutrition support service to help to reduce morbidity and mortality in patients.

Availability of nutrition support service encourages a hospital to use more enteral than parenteral nutrition, cutting down the cost of treatment.

Nutrition support does not mean merely giving protein, calories and minerals but providing substrates for the biological process of healing. It involves screening all patients on arrival—barring those with simple problems—for malnutrition underlying their disease, planning therapy and monitoring progress, using standard guidelines.

The ideal team for providing nutrition support service is headed by a physician, who has a special interest in nutrition. The co-director of the team is always a dietitian with a clinical dietitian and a food service dietitian as assistants. The role of clinical dietitian is to take charge of nutritional assessment, enteral and parenteral feeding, cholesterol control, and diabetes education. The food services dietitian oversees the food services manager and others who deliver the diet other members of the team are a part-time nurse, who maintains peripheral and central vein catheters and feeding tubes and a pharmacist, especially at centres where total parenteral nutrition solutions are prepared.

Responsibilities of a Hospital Nutrition Support Service: All high risk patients (those over 50 years, critically ill or suffering from trauma, cancer or kidney failure) are seen by the nutrition support service for a baseline nutritional screening. Revolution is done by day 3, even if the initial assessment did not reveal malnutrition. Appropriate interventions (enteral or parenteral nutrition or combination of the two) are recommended.

Protocols for enteral and parenteral nutrition are setup by the NSS, approved by the hospital administration and the medical staff and implemented.

Where formal teaching rounds with the entire treating team are not possible, interesting cases are discussed at scheduled sessions. Periodic journal clubs, meetings, case discussions and other educational activities are carried out. Research in clinical and basic science.

Pre and Post-operative Diet: Good nutrition prior to and following surgery ensures fewer post-operative complications better wound healing, short convalescence and lower mortality. Chronic diseases increase the nutritional requirements.

Malnutrition can lead to weight loss, poor wound healing, decreased intestinal motility, anaemia, oedema or dehydration and the presence of ulcers. The circulating blood volume and the concentration of the serum proteins, haemoglobin and electrolytes may be reduced.

Following surgery or injury the need for nutrients is greatly increased as a result of loss of blood, plasma or pus from the wound surface, haemorrhage from the gastrointestinal or pulmonary tract, vomiting and fever. During immobilisation, loss of some nutrients such as protein is accelerated. A fairly simple operation often involves moderate deficiency in food intake for a few days following the operation.

Some nutrients may be supplied by parenteral fluids, but the full needs of the body usually are not met by that means alone. Adequate oral intake is often delayed for a considerable period following cardiac or gastrointestinal surgery. Metabolic losses are great and alternative methods of nutritional support are needed.

Pre-operative Nutritional Assessment: Serum albumin and transferrin levels, triceps skin-fold and delayed hypersensitivity skin testing measurements are done to predict risk of complications in individual undergoing gastrointestinal surgery.

The objectives in the dietary management of surgical conditions are:

1. to improve the pre-operative nutrition whenever the operation is not of an emergency nature.

2. to maintain correct nutrition after operation or injury as far as possible and
3. to avoid harm from injudicious choice of foods.

Protein: A satisfactory state of protein nutrition ensures:

(a) rapid wound healing

(b) increases the resistance to infection

(c) exerts a protective action upon the liver against the toxic effects of anaesthesia and

(d) reduces the possibility of oedema at the site of the wound.

The presence of oedema is a hindrance to wound healing and in operations on the gastrointestinal tract, may reduce motility thus leading to distension.

When protein is depleted in post-operative condition complications are increased. Protein catabolism is increased for several days immediately following surgery or injury, patients are characteristically in negative nitrogen balance even though the protein intake may be appreciable. The degree of negative balance can be reduced at higher intakes of protein and calories.

The level of protein to be used in pre-operative and post-operative diets depends on the previous state of nutrition, the nature of the operation and the extent of the post-operative losses. Intake of 1.0 to 1.5 g per kilogram or about 100 g of protein are necessary as a rule.

Energy: Without sufficient caloric intake tissue proteins cannot be synthesised. Excess metabolism of body fat may lead to acidosis, whereas depletion of the liver glucose may increase the likelihood of damage to the liver. With 2500 to 3000 kcal patients make progress.

Obesity delays healing. Whenever possible, it should be corrected. Rapid weight loss results in loss of lean body mass and should be avoided.

Minerals: Phosphorus and potassium are lost in proportion to the breakdown of body tissue. In addition, derangements of sodium and chloride metabolism may occur subsequent to vomiting, diarrhoea, perspiration, drainage, anorexia and diuresis or renal failure.

Iron-deficiency anaemia occurs in association with malabsorption or excessive blood loss. Diet alone is ineffective in correction of anaemia, but a liberal intake of protein and ascorbic acid, together with administration of iron salt is of value in convalescence. Transfusions are usually required to overcome severe reduction in haemoglobin level.

Fluids: The fluid balance may be upset prior to and following surgery owing to failure to ingest normal quantities of fluids and to increased losses from vomiting, exudates, haemorrhage, diuresis and fever. A patient should not be operated in a state of dehydration since the subsequent dangers of acidosis are great. When dehydration exists prior to operation, parenteral fluids are administered, if the patient is unable to ingest sufficient liquid by mouth.

Vitamins: Ascorbic acid is especially important for wound healing and should be provided in increased amounts prior to and following the surgery. Vitamin K is of concern to the surgeon since the failure to synthesise vitamin K in the small intestine, the inability to absorb it or the defect in conversion to prothrombin is likely to result in bleeding. Haemorrhage is especially likely to occur in patients who have diseases of the liver.

Pre-operative Diet: Patients who have lost much weight prior to surgery benefit considerably by ingesting a high-protein, high-calorie diet for even a week or two prior to surgery. The diet may be of liquid, soft or regular consistency depending upon the nature of the pathologic condition. Parenteral nutrition or semisynthetic fibre-free diets are sometimes used. In addition, the maintenance of metabolic equilibrium as in diabetes or other diseases must not be overlooked.

Foods which provide a maximum amount of nutrients in a minimum volume are essential. Small feedings at frequent intervals are likely to be better accepted than large meals which cannot be fully consumed.

For additional protein, milk beverages may be fortified with non-fat dry milk or commercial protein supplements. Fruit juices fortified with glucose or high carbohydrate food, increase carbohydrate intake and facilitate storage of glycogen. Butter incorporated into foods and light cream mixed with equal amounts of milk are also useful for increasing the calorie intake. One should remember that the excessive use of sugars and fats may cause nausea.

Food and fluids are generally allowed until midnight just preceding the day of operation, although a light breakfast may be given when the operation, is scheduled for afternoon and local anaesthesia is to be used. It is essential that the stomach be empty prior to administering the anaesthesia so as to reduce the incidence of vomiting and subsequent danger of aspiration of vomitus.

When an operation is to be performed on the gastrointestinal tract, a diet very low in residue may be given 2 to 3 days prior to operation. In acute abdominal conditions such as appendicitis and cholecystitis, no food is allowed by mouth until nausea, vomiting, pain and distension have passed in order to prevent the danger of peritonitis.

Post-operative Diet: Following minor surgery, liquids are often tolerated within a few hours and rapid progression to a normal diet is made. After major surgery, oral intake may be delayed for days. Fruits like papaya and sapota are given post-operatively to initiate bowel movement. Complete nutritional support are provided by conventional intravenous feedings, catheter jejunostomy, total parenteral nutrition, tube feedings or semisynthetic fibre free diets.

Immunonutrition: Immunonutrition refers to the addition of specific nutrients in greater than normal concentrations to modulate immune function, and thus improve clinical function. The immunonutrients comprise of the following.

Arginine: A conditionally essential amino acid when stress occurs, arginine is a source of nitrogen in nitric oxide. It stimulates secretion of anabolic hormones and enhances wound healing and cellular immunity.

Purine Nucleotides: These are precursors for RNA and DNA. Deficiency of purine nucleotides is associated with immune suppression and supplementation improves cell-mediated immunity.

Omega 3 Fatty Acids: Omega-3 fatty acids have an anti-inflammatory effect. They are especially useful in reducing the severity of sepsis. Omega-3 fatty acids are found in fish oils.

Glutamine: A non-essential amino acid present in the body, and becomes conditionally essential in the presence of severe stress and multiple organ failure. It is a major oxidative substrate and enhances the immune function essential in stressed states. It is a substrate for glutathione synthesis and has cyto protective effects in animal studies.

Commercial formulations containing immuno nutrition are now available in the country.

Immunonutrition in Burns: Today, the practice in most burn centres is to initiate feeding within the first 12-24 hours. Animal data infact, show amazing tolerance when the gut is fed within 6 hours. It is possible, that early feeding before oedema develops in the bowel and gastric wall will help to prevent the alteration in motility that occurs in burn patients after 24-48 hours. Vitamin C has been shown to reduce oedema at the burn site, probably due to its antioxidant effect and that administration of trace elements (copper, selenium and zinc) is associated with a decreased use of antibiotics and a shorter length of hospital stay.

Studies have shown that enteral nutrition is associated with speedy recovery and discharge from hospital, reducing the cost of illness both for the patient and the institution.

There is evidence that the preservation of gut mass, prevention of increased gut permeability to bacteria and other toxins, support of splanchnic perfusion and maintenance of the gut associated lymphoid tissue play a part in reducing septic morbidity compared with bowel rest and total parenteral nutrition.

4

Food Constituents

The Liquids

About sixty to seventy per cent of the total body weight of a person is water. Water is present inside the cells of tissues (the intracellular water) and outside the tissue cells (the extracellular water).

The extracellular water is present in the blood plasma, water in the space between the tissues is known as interstitial water and in lymph.

Water and electrolytes are essential constituents of cellular metabolism. It is not only a component of our body but also acts as a vehicle for chemical transport. Water and electrolytes must be supplied to the body regularly to maintain the cellular function and to regulate the excretion through kidneys, lungs and skin.

Necessity of Water: Water intake and loss must be balanced. Water is in jested through food and also as drinking water. Apart from this, a considerable amount of water is formed in the tissues through the oxidation of carbohydrates, fats and protein in their metabolism.

Water Intake and Loss from the Body

Water intake	*(Ml)*	*Daily excretion*	*(Ml)*
As fluids	1,500-1,750	Urine	1,200-1,500
Water in solid food	600-900	Perspiration	700-900
Water from oxidation of carbohydrate, protein, fat metabolism	300-350	Faeces	100-200
Total	2,400-3,000		2,400-3,000

Water is lost through urine, skin, lungs and faecal matter. In a normal healthy person the- water intake and water loss is more or less equal. The excretion through the kidneys and skin changes according to the climatic conditions. In a dry climate more perspiration and in the cold weather more urination takes place. If the fluid intake is less thirst occurs. When two per cent of the body weight is lost dehydration starts.

Water is absorbed quickly and it is circulated to the tissues. From the tissues water is passed on to the capillaries. The capillary walls and cell membranes separate the water compartments. Water can move in and out of all cells by osmosis. The osmotic pressure in the plasma proteins regulates the exchange of water from tissues to capillaries. Antidiuretic hormone of posterior pituitary, hypothalamus, electrolytes like Na and K and adrenal cortex hormone aldosterone regulate the water metabolism in the body.

Antidiuretic hormone of the posterior pituitary increases the reabsorption of water in kidney tubules by increasing the permeability of distal tubules and collecting tubules. This mechanism enables them to retain more water in the body.

Hypothalamus is having the thirst centre. In dehydration plasma is concentrated and osmoconcentration stiumulates the thirst centre of hypothalamus. This induces the person to be very thirsty. By taking water the water balance in the body is regulated.

Various solutes present in the body fluids also help to regulate water metabolism. Plasma proteins exert low osmotic pressure and this helps to maintain the water in the blood vessels. Aldosterone increases reabsorption of Na by the kidney.

About 170 litres of fluid pass through the glomeruli of the kidneys daily. A healthy person's kidneys excrete about 1.5 litres of urine from the above fluids. The urine formation is reduced if the intake of fluids is less. Then the urine becomes more concentrated with urea which forms the main bulk in urine.

The normal output of urea is 30 mg for which 750 ml of urine must be formed. If urine formation is less urea is retained in the body which is harmful to the body; when less urine is formed in summer season more sweat is formed. Excessive sweating occurs with increased external temperature and humidity, fever and exercise. Water is also lost through insensible perspiration.

Apart from maintaining the intracellular and extracellular water metabolism, water also takes part in a number of activities. All the vital functions of the body depend on the presence of proper water content. Digestion depends on its juices and all digestive juices are watery in nature. Urine, the excretory medium of our body, contains 97 per cent water. Water also helps in regulating body temperature. By evaporation of water, by respiration and through sweating the body temperature is regulated.

Requirement of Water: Water requirement of a person varies with climate, age, activity, dietetic habits and body build. Activity is an important factor which affects the water requirements. When activity is more intense more heat is produced and more water is needed to dissipate the heat produced.

The type of diet also influences the water requirement. A

high mineral content increases the water requirement. Consumption of sweets demands extra water for their dilution in the stomach.

Beverages like tea, coffee and cocoa are diuretics and they stimulate excretion of water. If excess water is injested the kidney responds to it after 13-50 minutes and the extra water is excreted within three hours. The water balance will be maintained by the excess urine output. The sensation of thirst is one means of meeting the body's need for water.

Dehydration: Water depletion occurs when water loss is not met by water intake. On water depletion changes occur in the body fluids. Thirst, weakness, loss of weight, dryness of the skin, mucous membranes, tachycardia, disorientation and delerium are the symptoms manifested by dehydration. If not corrected coma and death occur. The urine changes in its specific gravity and concentration.

The salt level is diminished in the urine and serum levels of sodium, urea and chlorides are elevated. Intercellular water is depleted to compensate the water content of the secretions. This results in cellular dehydration and vital organs fail to function. Plasma volume and blood volume are reduced which leads to reduced output and further cardiac failure.

Acid-base Balance: Acid-base balance refers to the regulation of the hydrogen ion concentration of body fluids. From metabolic processes acids are continuously produced and have to be eliminated. Lungs and kidneys are the major agents in carrying out this function in our body. Normally the body is maintained in neutrality. In some pathological conditions like diabetes mellitus or fasting for weight reduction and in renal failure the acid-base balance is affected. Acidosis occurs in this condition.

Carbonic acid is the main form of acid formed in the body from metabolism. Carbohydrates, protein and fats yields Co_2

and water during their metabolism. The hydrated form of carbon dioxide is carbonic acid. Intermediary products like lactic acid, pyruvic acid, keto acids, uric acid, sulphuric acid, phosphoric acid from phosphoprotein are acidic in nature.

Certain foods in the body produce alkaline ash as the cations (Na^+. K^+) will allow the body to retain more bicarbonate ions thus producing alkaline effect. The neutrality of the body is maintained through the body fluids and their circulation, the acid buffer system like bicarbonate carbonic acid through haemoglobin and its capacity to form carbamale, regulation in respiration and the final effort from kidney on emergencies.

Acidosis: It is a condition in which the hydrogen ion concentration is increased or there is an excessive loss of base. The ratio of bicarbonate to carbonic acid is less than 20:1 and the pH is below 7.35.

Alkaksis: This is a condition in which the hydrogen ion concentration is decreased or the base is increased. The ratio of bicarbonate to carbonic acid is greater than 20:1 and the pH is 7.45.

Necessity of Milk: Milk is commonly considered as a complete food as it contains all six of the essential foodstuffs. Milk protein is of excellent quality and it promotes growth and maintenance of body tissues. Therefore, milk is an ideal food for infants and children. But milk is very low in iron and ascorbic acid content Calcium and phosphorus levels in milk are very high. The ratio of calcium and phosphorus and the presence of vitamin A help its maximum utilisation. Milk is an excellent source of vitamin A. In it the cream is rich in vitamin A and if the same is removed as in slammed milk, its vitamin A content becomes low. Riboflavin, a B vitamin, is present in significant quantities in milk. If the milk is exposed, to light riboflavin is lost. Fats present in milk are readily digested. The carbohydrate of milk is lactose and it is hydrolysed into glucose and galactose.

Whole milk contains 4.7 per cent carbohydrate, 4 per cent fat, 3.3 per cent protein and 88 per cent water. Fats in the form of oleic, palmitic and stearic acids is present in milk. Milk protein is casein and whey protein has lactalbumin and lactoglobulin.

The Average Composition of Milk and Milk Products (gm\100 ml)

Milk product	*Protein (gm)*	*Fat (gm)*	*Calcium (mg)*	*VitA. (I.U.)*	*Vit.C. (mg.)*	*Calories (kcal.)*
Buffalo's milk	4.3	8.8	210	160	3	118
Cow's milk	3.2	4.1	149	150	2	67
Buttermilk	0.8	1.1	30	...	...	15
Skimmed milk	2.5	0.1	120	...	...	29
Condensed milk	9.1	8.4	280	430	...	328
Cheese	24.1	25.1	790	273	...	348
Goat's milk	3.7	5.6	170	60	2	84
Human milk	1.0	3.9	20	70	3	67
Curd	2.9	2.9	120	40	4	51

In developing countries, the per capita availability of milk is poor whereas in developed countries it is more.

Among these buffalo's milk has a higher proportion of total solids and fats than cow's milk. Casein coagulation is also high.

Goat's milk is rich in cream and poor in protein compared to buffalo's milk. Human milk is easily digestible because it becomes solid on reaching the stomach. Casein is the chief protein of milk. It is a phosphoprotein which contains protein and phosphoric acid. An enzyme rennin aggregates and precipitates casein and it is coagulated or curdled. Bacterial action also curdles milk. Casein is present in milk in the form of calcium caseinate. Rennin converts calcium caseinate into calcium and paracasein. Again it forms a smooth, soft clot of calcium paracaseinate which is non-dispersible. If milk is overheated it forms a weaker clot.

Dried milk also forms a weak clot. Milk also contains lactalbumin, a simple protein of the albumin class. Prolonged and direct heating of milk also forms a weak clot. Milk also contains lactalbumin, a simple protein of the albumin class. Prolonged and direct heating of milk helps it to settle down at the bottom and sides of the container in which milk is heated. Use of double cooker rectifies this. The souring of milk is due to the presence of lactose, the carbohydrate. Lactic acid-forming bacteria use lactose as food to produce lactic acid in milk.

Fat is present in emulsified form. Creamy layer of milk consists of fats. Milk contains potassium, sodium, calcium and magnesium chlorides and phosphates. On heating milk calcium precipitates on the sides of the container. Of all food materials milk is the best source of calcium and careless heating reduces its availability. Even though iron is present in minute quantities it is highly nutritional.

Processing of Milk

Pasteurisation of Milk: Milk is a favourable medium for bacterial growth. Louis Pasteur, the French scientist developed the heat treatment of milk for inactivating disease-causing micro-organisms.

Milk-borne diseases are tuberculosis, typhoid, hoof and mouth diseases, undulant fever, scarlet fever, septic, sore throat gastroenteritis and diphtheria. But pasteurisation destroys certain advantageous bacteria also. There are two methods of pasteurisation—the flash method and the holding method. In the flash method the milk is heated at a temperature of 160°F for 30 to 60 seconds and in the holding method the milk is heated to temperatures of 140° to 145° F and held there for at least 30 minutes. In both the methods, after the heat treatment, the milk is cooled quickly to about 43° F. Pasteurisation reduces the thiamine and ascorbic acid content whereas the nutritive content of other vitamins is not much affected.

Evaporated Milk: Evaporated milk is produced by evaporating 60 per cent of water from milk and the product is then sterilised and sealed in cans. To evaporate, the milk is heated at a high vacuum from 130° to 150* F. The concentrated product is homogenised to reduce the fat globules and then placed in cans, heated and sterilised at temperatures from 240° to 245° F for IS minutes. Evaporated milk is fortified with vitamin D.

Condensed Milk: Evaporated milk is incorporated with 40 per cent carbohydrate to form condensed milk. Sucrose or dextrose or a mixture of these two is added.

Dry Milk: Whole milk or defatted whole milk is used in its preparation. From milk most of the water is removed. There are two methods applied to form dry milk-spray process and roller process.

In the spray process fresh milk is sprayed under high pressure into a drying chamber where dry air is passed. The current of dry air evaporates water and the dried solids settle to the bottom of the chamber.

In roller process fresh milk is spread over the surface of hot revolving metal rolls. The water evaporates and milk solids remain in thin sheets. If the fat is removed from whole milk the keeping quality is more. Dried whole milk does not keep so long because the fats tend to become rancid.

Various Milk Products: Cheese, yoghurt or curd, buttermilk, low-fat milk, non-fat milk cream and butter are the common milk products used.

Cheese is mainly from casein of milk with fat and a little whey. To precipitate cheese lactic acid or rennin are used. For commercial cheese making rennin is used and for homemade cheese lactic acid is used. Cheese is rich in protein and fat and a considerable amount of lactose is also present.

In homemade cheese calcium is less but other minerals like phosphorus are there in both. Compared to milk vitamin A content is the same but thiamine and riboflavin are dissolved in the whey. Cheese is used to spread in items like butter as it contributes flavour. In vegetables, fish and meat preparations and in custards and souffles cheese can be used. Low temperature is suggested for cheese cookery.

Buttermilk is prepared from curd; it is churned and the butter removed. Yoghurt or curd is a custard-like product prepared by fermenting milk. In commercial production streptococcus thermophilus is added to milk to ferment it. Low-fat milk is commercially prepared for people who prefer low calories from milk with a fat flavour.

Non-fat milk is skimmed milk where the fat is removed entirely. Vitamin A and D are fortified in it. Creams with different levels of fats are commercially produced.

The fat content ranges from 18 to 40 per cent. Cream combined with sugar, stabilisers and emulsifiers is often sold in cans. Commercially, cream is prepared by using a cream separator which enables the separation of fat at different levels. Whipped cream is an air-in-water foam in which fat globules are piled up in the film and it is used in salad dressings and in desserts.

Milk is used in many preparations for its solvent capacity and its nutritive contribution. Milk is used as a dispersing medium, especially in puddings and for scum formation with its coagulating property. Vegetables with tannin or acid cause milk to curdle. Scum formation and curdling occur when the protein in milk becomes insoluble in heat or in the pH change. Use of low temperature and short heating periods prevent scum formation. Acidity of the content produces curdling.

Adding acid food towards the end and mixing it in low temperature reduce curdling.

Types of Milks in the Market: Homogenised milk, separated milk, toned milk, sterilised milk, separated milk (skimmed milk), recombined milk, filled milk and flavoured milk are the common fluid milks available in the market.

Homogenised Milk: In this, fat globules are broken up mechanically to less than one microne in diameter. It is easily digested but cream is not obtained on separation.

Separated Milk (Skimmed Milk): This is prepared by removing fat from milk in a cream separator. Very little fat is in it.

Toned Milk: Toned milk is prepared by mixing milk reconstituted from skimmed milk powder with buffalo milk which has 7.0 per cent fat.

Recombined Milk: It is a homogenised product prepared from milk fat, non-fat milk solids and water. It should be pasteurised.

Filled Milk: Filled milk is the homogenised product prepared from refined vegetable oil and non-fat milk solids and water.

Sterilised Milk: Cow or buffalo milk is sterilised in bottles by fixating to 15 minutes.

Flavoured Milk: Chocolate, coffee, cardamom or any other edible flavours, colour and to flavour milk.

Other Beverages: Beverages are materials used as drinks for the purpose of relieving thirst and introducing fluid in the body, nourishing the body and stimulating or soothing the individual. Beverages are classified into four groups according to the reactions they produce.

(1) *Refreshing* to relieve thirst

(a) Plain or carbonated water.

(b) Ginger ale and other bottled beverages.

(c) Fruit juices and fruit-ades.

(d) Iced tea or coffee.

(2) *Nourishing*

(a) Milk pasteurised, skimmed, evaporated, dried, malted, buttermilk, and chocolate and cocoa.

(b) Eggnog made with rum, whisky, brandy, fruit juices, coffee and chocolate.

(c) Fruit juices with egg-white or whole egg.

(d) Glucose lemonade or orangeade.

(e) Tube feeding feeds.

(3) *Stimulating*

(a) Eggnogs made with whisky, rum, brandy, tea or coffee, other alcoholic beverages.

(b) Tea or coffee.

(4) *Soothing*

(a) Warm milk.

(b) Hot tea.

Coffee and tea have no food value unless served with milk and sugar. Cocoa in itself will increase the food value of the milk which in used its preparation. Coffee, tea and cocoa, all have a certain amount of stimulating effect. Iced coffee, tea and cocoa are used all the year round between meals as a refreshing drink.

Milk, eggnog and other cold beverages made with milk or milk and egg are valuable for serving between meals when the protein content of the diet needs to be increased. Wines and other liquors which are used occasionally to provoke appetite, or to stimulate or flavour furnish additional calories.

Cold beverages may be used with meals or between meals. They are more appetising when chilled. The choice of the beverage will depend largely upon whether it is taken for its food value or merely as a refreshment. In any case a cold drink served between meals adds interest to the monotonous hours in the daily routine of a person.

Tea: Tea is obtained from the leaves and flowers of teabush, which grows in sub-tropical countries, such as China, India and Japan. The top tiny leaves and buds or flowers at the end of the shoot furnish the choicest and the most expensive brands designated as flowery pekoe. Orange pekoe is the most commonly marked tea which consists of the first and second leaves just below the buds.

The kind of tea is determined by its treatment. Subsequent to picking, the green tea undergoes no fermentation process and is light in colour and rich in tannin. Black tea is fermented to a dark colour and loses some of its tannin.

A good cup of tea will contain a high proportion of volatile oils and a minimum amount of tannin. Tannin is a stimulant and has approximately the same psychological effect as caffeine. Tea, which is allowed to stand too long in the container with the leaves, contains a considerable amount of tannin which causes the finished beverage to be bitter and unpalatable.

Coffee: Coffee is the beans of coffee plants which grow in tropical countries. It is tasteless in its green form. Roasting develops the volatile oils, specially caffenol which give the characteristic aroma and flavour to coffee. The typical individual brands depend upon the various types of coffee used and the length of the time of roasting. Bitterness is increased when combined with substances like chicory.

Coffee alone has no food value, but it does furnish a certain amount of stimulation of caffeine which raises the blood pressure and stimulates renal activity and momentarily masks fatigue. As a rule a cup of coffee contains from 1.5 to 2.5 grains of caffeine depending upon the strength , of the brew. Tannin, another alcoholite present, interferes with digestion.

The amount of tannin in coffee depends upon the method of preparation. Since it is soluble in hot water, the longer the coffee is brewed, the greater will be the tannin content.

Coffee beans are ground to increase the surface area for maximum extraction of the flavour. However, coffeole is rapidly lost, the loss being greater in freshly ground coffee. Ground coffee should be kept in a tightly covered tin in a cool place and should be used promptly after purchase.

Drip and vacuum coffee contains the least amount of tannin, caffeine and coffeole. Percolated coffee contains slightly more caffeine and less coffeole than drip and vacuum coffee. Boiled coffee should be strained. It contains more caffeine and tannin than either the drip or percolated coffee.

Instant coffee is made from pure coffee which has been pulverised. The extraction of caffeine and tannin is high, but the amount in the finished beverage will depend upon the amount used. Some instant coffees are mixed with malted cereals.

In decaffeinated coffee 75 per cent of caffeine is removed. It does not possess the stimulating property of coffee. Coffee substitutes are made from a combination of roasted grains and cereals.

Cocoa: Cocoa is the ground product of cocoa beans and often some of the fat has been removed. The cocoa beans are in fleshy pods. The well ripened pods are feathered and dried for 24 hours after which they are cut open and the beans are removed. The beans then undergo a series of processes like fermenting or sweating, removing the pulp, washing, drying and roasting. Chocolate and cocoa, unlike tea and coffee, have a high nutritive value.

The roasted beans, likewise the bitter chocolate, contain approximately 50 per cent of fat, about 80 per cent protein, 10 per cent starch, and 7 per cent sugar. Like tea and coffee, they also contain a stimulant and tannin.

The stimulant known as theobromine has approximately the same physiological effect, though not to such a marked degree. There is only a small amount of stimulant in one cup and the chief nutritive value of the drink lies not in the cocoa but in the amount of milk and sugar used in it.

Coconut Water: Tender coconut water is high in sugar and electrolytes like Na and Potassium.

Fruit Juices: Fruit juices in themselves are not only pleasantly refreshing but they are easily digested and especially useful in increasing the fluid intake. The citrus juices and tomatoes are excellent for the ascorbic acid content Fresh, frozen and canned juices may be served plain or flavoured.

They may be reinforced with egg white or whole egg. The following fruits include commonly used juices: (1) citrus—orange, lemon, grape fruit, lime; (2) Berries—blackberries, loganberries, raspberries, strawberries; (3) tree fruits—apple, apricot, peach, pear, plum, prune; (4) other fruits—grapes, pineapple; (5) vegetables—tomato; (6) Sugarcane juice, sweet toddy and coconut milk are some of them.

Milk: Milk is most of all liquid food since it has more nutritive value than any other beverage. Milk beverages add protein, calories, calcium and other nutrients to the diet. The milk flavour of food permits its use in a great number of ways. It may be plain, malted, or acidulated. It may be reinforced with egg and sugar depending upon the type of nutrient needed and it may be flavoured with chocolate and other flavouring extracts.

Egg: Eggs are specially useful for increasing the protein content of beverages. Eggnog in which whisky or other alcoholic beverages are used is the most important stimulating beverage. Eggnog made with strong coffee is only mildly stimulating but lends variety to the general liquid diets. Eggnogs contain egg, cream, and possibly milk and sugar in addition to fruit juice.

Sugar: Sugarcane, fructose and lactose may be used to sweeten beverages and to increase the caloric value.

Bask Rules in the Preparation of Tea and Coffee: The best tea produces a pale coloured infusion. Freshly infused tea has good flavour and it is harmless for digestion. Reheating or overheating of tea results in continued infusion and it extracts more tannin which gives a bitter taste Id tea. Moreover, this is often harmful to the digestive system.

The quality of tea is an important factor in giving good taste to the product. Since tea leaves absorb water, they must be stored in airtight containers.

Water used for the preparation of tea also has the key role in producing good result. Water should be freshly drawn before it is used for tea. Previously drawn water and previously boiled water should not be used for preparing tea as it makes the tea insipid. To get the best flavour water should not be boiled before or after tea leaves are added. As soon as the tea leaves are added to boiling water it circulates from the bottom to the top and from the top to the bottom.

To prepare good tea, one teaspoonful of tea leaves for 1 cup is ideal. For one litre of water 15 gms of tea leaves are enough. It is always better to measure tea leaves to get good result. The vessel in which tea is being prepared must be warmed before making tea. Instead of adding tea leaves to water, pour boiling water into a wanned pot in which tea leaves are added. Do not disturb this for five minutes. Let the leaves infuse. If it is kept for more than five minutes, it gives a bitter taste and less then five minutes yields a weak tea.

Milk and sugar can be added according to individual taste. To get a good cup of tea, it must be served hot.

Coffee: There are different methods of preparing coffee. The pot method, the filter method, the urn brewing method and the vacuum brewing method are the common methods.

Coffee powder must be fresh to get the best flavour and taste. The quantity of coffee powder used must be just enough. For one cup of coffee 8.5 gms or 1/2 teaspoon of brewing coffee is required. If instant coffee is used one teaspoon is enough to make 1 cup of coffee.

Freshly drawn water must be used to prepare coffee.

Coffee is generally prepared by any one of the following:

1. The pot method.
2. Filter method.
3. The percolator method.
4. Instant coffee preparation.

The Pot Method: In this method an earthenware pot or jug is used to prepare coffee. Warm the earthenware pot or jug, put in 3 level spoons or 11/2 heaped teaspoons of fresh coffee powder (coarse grind) for an 8-ounce cup. Pour boiled water over the powder and stir. Cover the vessel and let it stand for 5 minutes. Strain the coffee through a fine meshed sieve or cloth. Add milk and sugar to taste.

Filter Method: Stainless steel or brass filters can be used to filter coffee in this method. If brass filter is used it should be tinned properly to get good results. Put 2 to 3 level teaspoons coffee powder for 8 ounces water. Press the plunger down lightly over the powder and pour boiling water over the plunger in a circular motion. Allow to stand for 5 to 7 minutes. Pour out coffee from the lower vessel and add milk and sugar to taste.

The Percolator Method: Keep coffee powder in the middle of the percolator and pour boiling water through it to the upper section. Coffee is collected from the bottom section. It takes thirty minutes to percolate down. Vacuum coffee makers are also available in some places.

Instant Coffee: Strong brew is made by percolation of roasted and ground coffee which is sprayed into fine atomised particles from the top of a chamber into a mass of very hot air. The flavour is retained through highly developed technology. It is quickly packed into airtight bottles.

Varying degrees of chicory is mixed with coffee powder to give slight bitterness and colour to brew. Now mixtures of coffee soluble with solids and sugar in the form of thick pastes are also marketed. By adding hot water it can be used instantly.

Alcohols: Beer, ale, toddy, wine whisky, gin, brandy, rum, vodka, arrack and cider are some of the alcoholic beverages. Beer is made by fermenting barley malt extract with yeast and ale is the same thing with more stimulation. Toddy is from fermenting sap of palm, coconut and palmyra or date palm.

Wine: Wine is prepared by the fermentation of grapes with yeast and cider is apple wine.

Whisky: It is an alcoholic distillate from a fermented mash of grains.

Gin: It is an artificial drink from diluted pure ethyl alcohol.

Brandy: This is made from fruit juices like grapes and apples.

Use of Non-veg

Flesh of all edible animals including mammals and poultry is termed as meat. Meats of cattle origin, beef and veal, of hog origin, pork, sheep, lamb, mutton, flesh of partridge, duck and game birds are used commonly as food.

Meat: Muscle of meat is composed of 75 per cent water, 20 per cent protein and 5 per cent fat, carbohydrate and minerals. The composition of meat varies with its sources, season of the year and pH of the meat. Protein comprises about four-fifths of the solid material. True fats, fat-like substances called phospholipids, inorganic substances and vitamins are present in fats.

Veal is the meat from cattle slaughtered three to fourteen weeks after birth. Calf is from fourteen to fifty-two weeks old animal. Beef is from cattle over one year old. A heifer is a female cattle that has never borne a calf and a cow is a female cattle that has borne a calf. Meat has skeletal or muscular cuts and organ cuts. Skeletal or muscular cut has muscle, tissue, çonnective tissue, bone fats and nerve tissues. Muscle tissue and connective tissue supply the qualities of meat in cookery.

Muscle structure is made up of bundles of muscle fibres called fasciculi. Connective tissues are scattered throughout the muscles where they play a connecting and supporting role. White fibrous and yellow elastic fibres with proteins like collagen and elastin are present in all muscles. Collagen contributes strength and elastin supplies elastic quality to meat Collagen reacts with water in hydrolysis to produce gelatin and brings about a tenderising effect to the meat. But no sack change is brought in elastin and if elastin is more in muscle fibres meat will be hard even after cooking. This can be remedied by chopping or grinding the elastin fibres.

Fats in meats are distributed between fasciculi. Fat is laid as a protective layer around the organs. The fat deposited with the muscle structure is known as marbling. Muscle haemoglobin is present in small amounts. Muscle cells of living animals contain glycogen which is otherwise known as animal starch.

But immediately after the animal is dead glycogen is converted into lactic acid. The vitamin content of different meats varies. The B vitamins—thiamine, riboflavin and niacin—occur in significant amounts in all meats. The meat of hog origin is rich in thiamine liver is rich in vitamin A. Iron-containing substances and B vitamins are more in organ meats.

Rigor Mortis: After the animal is slaughtered certain changes occur in the meat and it stiffens. This process is known as rigor mortis. On slaughtering the animal adenosine triphosphate is synthesised from glycogen and lactic acid, ammonia and other products are formed from it.

Myosin and actomyosin are found in contracted muscles and stiffening occurs to it. If the slaughtered animal has a high glycogen level, it produces a low pH and the quality of meat will be good with less rigor mortis.

Fasting, exercise, nervous exhaustion or insulin injection cause low glycogen level and the carcass has a high pH. Undesirable dark colour and poor flavour and texture of meat result from this. Slow and long slaughter produce more adrenaline in the animals and this results in sticky, gummy and dark coloured meat. Rigor mortis reaches its peak within 24 hours and then the muscle softens.

Quality and Grading: Appearance, tenderness, juiciness and flavour of the meat decide the quality of meat. Good quality flesh must be firm, the fasciculi should be small and velvety in texture, with characteristic colour of the meat-bright cherry red for beef, pinkish-red for lamb, greyish pink for veal and pork; bone must be small, red and spongy in appearance and the fat must be white and firm. To grade the meat quality grades are given and they are prime, choice, good, commercial, utility and cutter and conner.

For lambs and veal cull is the term used instead of utility. For pork grades of 1,2, and 3 are used. Meat of top quality is more tender than lower quality grade. Tenderness of meat increases palatability and it varies according to the cuts of meat. The most tender cuts are from backbone and ribs. Less tender cuts are from the shank, shoulder, neck and under parts of the body. Under part cuts of beef are flank, plate and brisket. The same cuts from lamb and veal are known as breast.

Meat Cookery: Palatability, appearance and flavour are enhanced by cooking meat. Cooking also makes meat more tender and digestible. There are two methods generally applied to meat cookery—dry heat and moist heat methods. In all methods of cooking the interior of the meat changes. Heat penetration, substances added to meat and composition of the meat affect these changes. Mainly the method adopted for cooking depends upon the cuts of meat.

Less tender cuts contain more connective tissues than tender cuts and so moist heat methods convert collagen to gelatin and it has a tenderising effect upon tissues. Roasting and broiling are the methods used to cook by dry heat methods and these methods can be applied to tender cuts of meat. For less tender cuts moist heat methods are used. Moderate temperature is recommended for this method. Water, juices and fluids are used for most heat methods. The addition of sodium chloride or sodium and potassium phosphate has been reported to increase the tenderness of meat.

Dry Heat Methods: *Roasting:* Roasting is cooking by means of radiant heat. Meat is roasted either in the open or in a closed tin in an oven. In covered roasting evaporation is diminished, and it cooks faster too. The water runs out of the meat instead of evaporating from the surface. An enclosed roast loses more weight and water than the open roast. Cooking to rare stage only increases the juiciness and flavour. If the pieces of meat lack fat, a thin layer of fat on top will be very effective. Water should not be added to oven roast.

Broiling and Pan Broiling: Slices from tender cuts of meat can be cooked rapidly without adding moisture or fat by radiant heat from a broiler or in an uncovered pan. The time required depends on the stage of cooking required and the size and shape of the pieces. This method gives a uniform result. The temperature should be higher than in roasting. It can be regulated either by changing the heat input or by changing the distance of meat from the heat source. This method is suitable for cooking tender cuts of beef and lamb or ground meat on the pattern of ham, bacon and sausages.

Frying: Meat is fried when it is partly or completely immersed in hot fat during cooking. When the meat is completely immersed the method is called deep-fat-frying. When a small amount of fat is added it is called pan-frying or sauteing. The fat should be kept below smoking point.

A temperature of 350° F for pre-cooked meat and 300° F for raw meat is usually satisfactory. This method is suited to lean meats such as liver or cooked steak.

Cooking by Moist Heat: The cooking medium is steam of water.

Braising and Pressure Cooking: Meat may be cooked in steam with or without pressure. When the meat is first browned in its own fat or a small amount of added fat and allowed to cook in its own steam without pressure the method is known as braising. Meat braised as a large piece is called a pot roast and that as small pieces, a fricassee. Braising is preferable to boiling for fresh pork, because it produces less hardening and drying of fibres when meat is cooked to a safe stage of cooking.

Stewing or Simmering: Meat may be cooked by immersing it in water. To minimise the hardening effect of high temperature simmering is preferable to boiling. Vegetables are added to stews to enhance the flavour.

Generally shrinkage of meat cooked by moist heat method is greater than that in dry heat methods.

Tenderising Meat: Papain, an enzyme derived from papaya, tenderises beef. Solution of papain is often injected into the beef shortly before slaughter. Meat of old animals is tough and stringy and usually this meat requires tenderising. Tougher cuts of meat require a long, slow cooking period. Prolonged cooking changes collagen into gelatin in moist heating. Dry heat hardens it. Acid hastens this conversion of collagen into gelatin; tomato juice and vinegar tenderise meat. High temperature gives a tasteless dry-product-like quality to tough meat. Cutting the tough meat into smaller pieces enhances the flavour of meat. Mining, grinding or pounding of meat helps to break the muscle fibres.

Vitamin retention is more in dry heat methods. Roasting or broiling retains 65-70 per cent of thiamine whereas moist heat retains only 25 per cent of thiamine.

Gelatin: Gelatin is a protein obtained from collagen on hydrolysis. Connective tissues of cartilage, tendons, bones, skins and hoofs are used to prepare gelatin. After preliminary treatments a dispersion of collagen in water is obtained. And this dispersion is boiled for several hours. Water is evaporated and gelatin is obtained from collagen.

From the nutritional point of view gelatin is inferior because it is an incomplete protein. The property of gel formation is used in cookery. In desserts gelatin is used to provide a spongy appearance to the product.

Processing of Meat: Canned meat, cured meat, smoked meat, dehydrated meat and sausages are the common processed form of meat.

Canned-meat: The meat is cooked and filled in the can along with the gravy, leaving a little head space.

Cured-meat: Meat is pickled by the use of curing agents like sodium chloride, sodium nitrate and vinegar.

Smoked-meat: Smoking helps to preserve the meat and develop flavours in it The temperature and period of smoking vary with the type of meat

Dehydrated-meat: The meat is cut into pieces and cooked in steam for 30 minutes at 10 lb pressure. It is passed through a meat chopper and the chopped meat is dried in a drier.

Sausages: Cooked chopped meat is ground to a paste with seasonings and packed casings. It is available as fresh, cooked, smoked, cooked and smoked, dry, and semi-dry.

Fish: Fish is a very good source of calcium, protein, vitamins A, D and iodine. There are two main classes into which fish may be grouped as vertebrate fish or fin fish and shell fish. Cod, mackerel and salmon have backbones and fins, whereas a shellfish is more or less completely covered with a shell. Oysters, lobsters and crabs are examples of shellfish.

Protein is the main component in fish and its content ranges from 15 to 20 per cent by weight of flesh. The fat content in fish varies from 1 per cent to 20 per cent or more. Hilsa has a fat content of about 19.4 per cent. The fat content of shellfish is less than 5 per cent. Since the fats in fish are unsaturated fats even in therapeutic diets this can be recommended. The flesh of fish is devoid of carbohydrates except in the case of shellfish. The mineral content of fish is variable; 1 to 1.5 per cent of the flesh of fish is with mineral elements. Small fish which can be consumed with their bones are rich in calcium. Oysters and shrimps are rich in calcium. Most fishes are not very good sources of iron except oysters. Sea fish are high in iodine content. Oysters are also rich in copper. The flesh of most fish contains fair amounts of thiamine and riboflavin and some amounts of niacin. Fatty fish are rich in vitamins A and D. Fish livers are rich sources of these vitamins.

Freshness of Fish: Fish deteriorates quickly and so requires careful handling. The quality of freshness of fish is indicated by bright red gills, clear eyes, bright and slightly bulging but firm flesh when pressed with a finger, and elasticity. The fresh odour of fish suggests its freshness. Suggestion of putrefaction is easy to detect from its foul smell. The tail of a fresh fish is stiff.

Spoilage of Fish: Fish is highly perishable. The spoilage of fish is due to microbiological, physiological and biochemical changes.

Processing of Fish: Fish protein concentrates, fish meat, smoked and salted fish are the common forms of processed fish. Fish protein concentrate is used for supplementing diets.

Fish meals is not suitable for human consumption. It is used for poultry and for animal ration.

Preservation of Fish: Cold storage, freezing, canning and drying are the common methods of preservation of fish. Salting before drying is essential.

The Eggs: Eggs are used in many preparations due to the variety they provide to food. Eggs are important ingredients of recipes for breakfast, lunch or dinner for any person. Nutritive importance, flavour, pleasing colour, thickening power, foam-forming ability, emulsifying capacity, stabilising capacity and the binding property of eggs are utilised in menu planning.

An egg has two parts—outer egg-white and inner egg yolk. The egg is held in position in the shell by two membranes which lie close together. About 63 per cent by weight of an egg is made of egg-white. It has three layers—an outer portion of thin egg-white, a thick jelly-like layer and a thin portion which surrounds the yolk. The central part of the egg is the egg yolk, and the yolk is surrounded by a thin covering called vitellin membrane.

The egg-white is rich in protein. Albumin, globulin and mucin are the proteins in egg-white. It is easily coagulated by heat and it provides the jelly-like character of thick egg-white. Minute quantities of sodium, potassium, magnesium, calcium, clorides and phosphates and riboflavin are present in the egg-white.

The egg yolk contains lipovitellin as the proteins and it is phosphoprotein, and it resembles the casein of milk. The egg yolk is rich in fat, that is, about one-third of the yolk is fat, especially of cholesterol and lecithin variety. The fats in egg are finely dispersed and are in emulsified forms. The mineral content is more in egg yolk compared to the white. Egg yolk is an important source of iron and it is also rich in sodium, potassium, calcium and magnesium.

Lipovitellin and lipovitellinin are the lipo proteins in egg yolk which act as emulsifying agents. Egg yolk is rich in vitamins, especially thiamine riboflavin, vitamins A and D and their content in egg depends upon the hen's diet.

The whole egg is a rich source of all nutrients except vitamin C. The protein in egg is considered as a complete protein because of its high biological value and digestibility. It is often mentioned as reference protein. But a raw egg-white contains a protein known as avidin which renders the vitamins biotin unavailable to the body. Duck egg-white contains another substance known as trypsin-inhibitor which inhibits the action of trypsin on protein. Heating destroys avidin and trypsin-inhibitors.

Role of Egg: Egg protein coagulates on heating. The coagulum formed is a smooth and jelly-like mass. It supplies a thickening effect to the products. If the temperature is high, the coagulum is toughened and it shrinks and water comes out of the product. This is called syneresis and it occurs in custards and scrambled egg preparations.

Egg is cooked in its shell, scrambled, poached, fried or used in the preparations of deserts, and cakes and other confectioneries. Egg is used in food mixtures to contribute flavour and colour to the products. Cakes and puddings owe their colour and flavour to eggs. In the preparation of mayonnaise the protein of egg stabilises the oil in water dispersion.

The protein of egg stabilises the air in liquid dispersion. Egg is also used for thickening various desserts and creams and patties. Since egg is highly perishable the quality has to be judged by internal appearance flavour and external shell appearance.

Candling: Candling is used as a method to detect the interior quality of egg. In a closed box with a small opening a lighted candle has to be placed. The depth of air cell can be observed by placing the egg in front of the opening. In a fresh egg the white is stationary and slight movements of yolk are permitted by the white. Yolk is in the centre and it does not move nearer to the shell.· In a poor quality egg the yolk moves. The shell of good quality egg is firm and not stained or cracked.

Formation of Dark Greenish Discolouration: When egg is boiled for more than 15 minutes a dark greenish colour may be formed on the surface of the egg yolk. This green colour is due to the formation of FeS on the surface of the yolk. Yolk contains more iron and the white has more sulphur in its amino acids. On boiling it yields H_2S. The H_2S combines with the iron of the yolk and it gives the dark green colour in the middle. In high temperature this change takes place immediately.

Use of Veg

A fruit is the edible fleshy seed-bearing part of a plant or tree. Fruits are favoured by all due to their attractive appearance, appealing flavour and pleasing odour. Apart from their nutritive value they are also appreciated for their appetising effect. Fruits contain starch. On ripening the starch is converted into fructose and glucose and these sugars contribute to the sweetness of fruits.

Water is the chief constituent of fruits. About 75 to 95 per cent of the total weight is from water. Fruits as a whole are rich sources of vitamin C and vitamin A and minerals like sodium, potassium, magnesium, fibre and cellulose. Characteristic flavours of fruits are due to the presence of organic acids in them. Citric acid and malic acid are the acids present in fruits. Grapes contain tartaric acid. Acid content of various fruits varies.

Citrus fruits are the best sources of vitamin C. Dried fruits like raisins, dates and figs are rich in iron. Calcium is present in citrus fruits. Banana contributes a very high amount of carbohydrates and about 23 per cent of it is sugar.

Jackfruits, figs and cherries are also very good sources of carbohydrates. For vitamin C guava is a cheap source. About 300 mg vitamin C is present in 100 gms of guava. Cashew fruit is also an excellent source of vitamin C. Lime, mango, papaya and orange are other sources of vitamin C. Fruits have alkaline reaction in the body.

Fruits: Fruits with different colours have carotenoids, anthocyanins and flavanols in them. Fruits also contain enzymes like ascorbic oxidase and phenol oxidase. The vitamin A value of fruits is in the carotene in yellow fruits. Organic acids and salts are present in fruits. Dried fruits like dates and raisins are sources of iron. Potassium is the most important salt in fruits. Potassium is present in citrus fruits as citrates and in grapes as tartarate.

Pleasing colours of the fruits are due to their pigment content. In ripe fruits carotenokl pigments are present in different forms. Red, purple and blue shades of fruits are due to the presence of anthocyanine pigments in them.

Tannins and Browning Reaction: Tannins are present in almost all fruits. They contribute bitter taste and astringent quality and discolouration of cut surfaces of fruits on exposure to air. When certain fruits are pared a brown colour appears on their surface which becomes darker as the fruit is allowed to stand. Apple, banana, plum, guava and peach show this change. Tannins and flavone pigments give brown coloured pigments on oxidation.

Oxidising agents present in fruit cells catalyse the reaction. If the fruits are cut with an iron knife, tannins and flavones react with iron salts forming a greenish purple compound which discolours the fruits. Acidic juices of fruits like lime, grape or pineapple are sprinkled on the fruits to prevent browning action.

Enzymatic browning occurs only when the fruits or vegetables are cut into pieces. The natural phenolic compounds present in tissues of plants do not come into contact with the phenoloxidase which is also present in some tissues. Practical methods for preventing enzymatic browning are to change the pH and the use of anti-oxidants which retard oxidation. By using the acidic juices of fruits the pH can be brought down.

Non-enzymatic browning occurs due to the formation of brown condensation products from reducing sugars and amino acids on heating. To prevent non-enzymatic browning storing the food materials at a low temperature, between 0° and 5°C, and addition of sulphur dioxide in the case of dehydrated vegetables are recommended. Browning reaction affects the nutritive value of food, especially the lysine content of it. Loss of nutrients of fruits on cooking is negligible.

Prevention of Browning: Thermal inactivation of oxidase enzymes, elimination of O_2 from the systems, change of pH to prevent enzyme action, and use of antioxidants are the measures to prevent enzymatic browning.

For non-enzymatic browning low temperature (0°-5°C), addition of SO_2 and keeping the moisture content low are some measures.

Vegetables: Edible parts of plants like bulbs, tubes, roots, seeds, fruits, stems, leaves or flowers are considered as vegetables. Vegetables as a whole are important sources of minerals and vitamins and certain vegetables have starch, cellulose and water. In meal planning and for dietary calculation vegetables occupy a prominent place.

To check their sufficiency and also for convenience, vegetables are grouped into green leafy vegetables, and roots and tubers and other vegetables. Legumes like double beans, french beans, cluster beans and peas are also included into other vegetables.

Vegetables are the best sources for calcium, iron, sodium, chlorine, cobalt, copper, magnesium, manganese, phosphorus and potassium. Carotenes, the precursors of vitamin A and ascorbic acid are present in abundant quantities in leafy and yellow vegetables. Vegetables are poor sources of calories except in the case of roots and tubers. Raw legumes and beans are good sources of proteins. Vegetables as a group supply a high amount of water and the mineral elements present in them contribute the alkaline substances in the body. This enables to maintain the acid-base balance of the body.

Cabbage, cauliflower, coriander, leaves, lettuce, spinach, amaranth and drumstick leaves are the common leafy vegetables. Of all the vegetables green leafy vegetables are the richest in carotene and the carotene content depends upon the green colour. If the colour is deep the carotene is more in it. Green leafy vegetables are fair sources of proteins and good sources of folk; acid, ascorbic acid and calcium. Some leafy vegetables contain oxalic acid which interferes with the absorption of calcium in the diet. Leafy vegetables act as a buffer and maintain the Proper alkalinity of the blood by balancing the acidity of acid-producing foods like meat.

Roots and Tubers: Roots and tubers are good sources of starch and fair sources of protein, vitamin B and vitamin C. Roots and tubers include tapioca, potato, sweet potato, carrots and beetroots. Carrots are rich sources of Vitamin A. The starch in roots is easily digestible. Potato, a root vegetable, is a fair source of proteins, ascorbic acid, and B vitamins. Sweet potato is a poor source of proteins but it is a fair source of B vitamins and ascorbic acid. The yellow variety of sweet potato is a rich source of pro-vitamin A or carotene.

Other Vegetables: Other vegetables are good sources of minerals, cellulose and water. They are also good sources of vitamins. Common vegetables like ladyfingers, gourds, brinjals, pumpkins, green peas and all other vegetables fall in group Tomato is also considered as another vegetable. Green peas and yellow pumpkin are fair sources of carotene.

Cooking of Vegetables: Cooking affects the colour, texture and nutritive value of vegetables. Cooking methods can destroy the good qualities if not handled properly. Common methods of cooking are boiling, steaming, baking, frying and pressure cooking. Before applying these methods the chemical structure of vegetables must be considered.

Vegetables contain different chemical compounds like pigments, tannin, enzymes and flavouring substances. Common pigments in vegetables are chlorophyll, carotinoid, flavones and anthocyanins. On heating these pigments undergo certain chemical changes which affect the colour of the finished product. Vegetables also lose organic acids and nutrients in cooking water. Acid or alkali content, and the cooking media also affect the appearance and quality of the product

Chlorophyll: Chlorophyll is the green pigment in leaves which is not readily soluble in water. In an acid medium chlorophyll turns an olive green and in alkali it becomes bright green. But the use of alkali to retain the green colour in salad vegetables destroys vitamin C and thiamine. Cooking in copper vessels turns chlorophyll into deep bluish green colour. Addition of sodium carbonate affects the texture and the cellulose becomes mushy.

Cooking with a lid and in minimum time retains the colour. The Carotenoid is yellow orange in colour and insoluble in water. It is soluble in ether and acetone. Carotenoids are present in different forms like carotenes, Ly copence, xanthophylls, cryptoxanthin and citroxanthin. Carotenoids are present in yellow vegetables like carrots, sweet potatoes, tomatoes, and xanthophyll in all vegetables. Cryptoxanthin is present in papaya, orange peel and yellow maize. Carotenoids are unchanged in acid or alkali and so vegetables with carotenoid pigments may be cooked with a cover. Such vegetables can be cooked by any method as the carotenoids remain unaffected.

Flavanoids: Flavones and flavonols are together known as flavanoids. Anthocyanins, anthoxanthins, leucoanthoxanthins and catechins are together known as flavonoids.

Anthocyanins are reddish water soluble pigments occurring in fruits and vegetables. They are destroyed by heat and oxidation. Passion fruit, apple, cherry, plum, grapes, blackberries are some of the fruits with anthocyanin in them. Beetroot has anthocyanin as its pigment Anthocyanin is water soluble and in acid medium it turns blue. In some vegetables flavones and anthocyanin are combined together which gives a green colour on cooking in alkaline medium.

Another chemical substance present in plants is tannin. Tannin and ferric salts form a greenish purple compound. On oxidation this turns brown which imparts a brown colour. Certain vegetables show discolouration if an iron knife is used to cut it. If vegetables rich in tannin are cooked in milk, the milk curdles.

Effect of Cooking: At a high temperature the cellulose is disintegrated and protopectin, a non-dispensable substance in water, undergoes hydrolysis and makes it dispersible. This changes the structural framework of the vegetables and makes it soft. If it is overcooked the vegetable becomes mushy and unpalatable. Some vegetables lose their natural flavour on cooking. Strong flavoured vegetables like onion, cabbage and cauliflower must be cooked by suitable method either to retain or to lose the flavour.

If a mild flavour is desired the vegetables can be cut into small pieces and cooked in a small amount of boiling water in a covered pan. If vegetables are in large pieces, large quantity of water can be used in an open pan. The plant acids are dissolved in water and the open lid allows the acids to escape.

Green vegetables turn brown because the plant acids and cholorophyll are disintegrated on cooking. Cooking time must be reduced so as to prevent their contact. Use of pressure cooker, large quantity of boiling water to cook the vegetables and use of an uncovered part are some of the measures that can be used to reduce the colour change.

Retention of Nutrients: Vegetables are rich in mineral salts and vitamins and most of them are water soluble. Some of them which are not soluble in water are destroyed by heat or by oxidation. If cooking water is thrown away considerable loss of mineral salts, especially of sodium, potassium and chlorine due to leaching occurs. To avoid this either use minimum , amount of water or the excess water after cooking can be used for other preparations.

Cut the vegetables as big as possible to reduce the cooking lose bring the cooking water to boiling before inserting vegetables. Cook just enough. Washing vegetables after cutting enables considerable loss of minerals in the water. About 40 per cent minerals and thiamine are lost if the rice is washed twice or thrice -before cooking. Washing once and utilising the rice water are better methods for retaining nutrients. Absorption method of cooking can be adopted for rice cooking.

Vitamin A or carotene is not affected when food is cooked in water. But shallow frying or roasting brings considerable loss of these vitamins. If sodium bicarbonate is added especially for dals or pulses most of the thiamine is destroyed. Roots like potato, sweet potato and other roots which can be cooked with their skin must be cooked so as to prevent leaching out of nutrients. They can be boiled first and later cut into pieces. Vitamin C is the most unstable vitamin among water soluble vitamins. Washing, cutting with an iron knife, cutting into smaller pieces and prolonged cooking lead to oxidation of vitamin C content of food items. If tamarind or lime, which is high in acidity, is added to cooking water it has a preservative effect on this vitamin. Since vitamin C is leached the least possible amount of water must be used and the cooking done in a covered vessel.

Cooking just before consumption is another method to preserve vitamin C content in the foods. Use of pressure cooker minimises the loss of nutrients. Consumption of raw vegetables provides all nutrients without any loss. If boiling is the method used to cook vegetables put them only into boiling water and bring the cooking water again to boiling temperature. For strong flavoured vegetables use little boiling water and let them remain uncovered. For other vegetables smaller amounts of water and tightly covered untensils are recommended. Avoid the use of baking soda.

The Losses of Vitamins in Some Vegetables during Cooking

Name of vegetable	*Method of cooking*	*Tiamine*	*Loss of Riboflavin*	*Vitamin C*
Beans	Boiling	29%	20%	42%
Cabbage	"	32%	18%	45%
Carrots	"	20%	10%	30%
Potato	"	25%	9%	23%
Peas	"	28%	17%	40%
Spinach	"	15%	10%	28%
Sweet potato	"	20%	13%	24%
Beans	Steaming	22%	14%	32%
Cabbage	"	18%	9%	38%
Carrots	"	15%	10%	25%
Peas	"	24%	11%	30%
Potato	"	20%	12%	25%
Spinach	"	16%	12%	25%

Use of Cereals

Dried edible seeds of certain plants are called pulses. Split pulses are known as dals. Pulses are good sources of proteins and B vitamins. Even though vitamin C is not present in pulses, sprouting of pulses makes it a very good source of vitamin C. Pulses are the best supplements for cereals.

Generally pulses contain about 20 to 28 per cent of protein but they are inferior in quality compared to animal foods. Methionine, the limiting amino acid in cereal proteins, is present in good amount in pulses. Thus cereal pulse combination improves the nutritional quality of the diet. This combination is the most economic way to provide protein.

Pulses contain fair amounts of minerals like calcium and iron. About 8 to 10 mg of iron and about 100-200 mg of calcium are present in 100 gms of pulses. Pulses contain fair amounts of thiamine, riboflavin and niacin. Sprouting of pulses improves the carotene content and vitamin C.

Pulses: Green gram, red gram, bengal gram, black gram, cow gram, horse gram, lentils, peas and soyabeans are the common pulses used in our cookery. All these pulses are good in protein, carbohydrate and vitamin B and fair in mineral content. But vitamin C is not present in them. On sprouting this deficiency can be rectified.

To sprout, cleaned pulse is soaked for 24 hours and then it is spread on a muslin cloth, sprinkling water occasionally on it. After 24 hours tiny sprouts appear on the surface. These sprouts are very rich in vitamin C and carotene.

Sprouted Pulses are very tender and cooking is not necessary. They can be used as raw salad by sprinkling lime juice and salt or as raitas by adding curd and salt to it. As cooking destroys vitamin C it is better to use them as a raw item.

Vitamin 'C' Content of Sprouted Legumes

Period of Germination	*Vitamin C*
Dry legumes	Trace
After 24 hrs germination	7-12 mgms/100 gm
After 48 hrs germination	11-18 mgm/100 gm
After 72 hrs germination	13-20 mgm

If pulses are not cooked properly they contain much undigestible materials. But if pulses are used as flours they are completely digested and absorbed. In raw pulses trypsin inhibiter is present and it is anti-digestive in nature. Among the various pulses green gram is digested best.

If more than 100 gms of pulse is consumed in a day it is not properly digested and gas is produced. If diarrhoea or other stomach complaints are present gas is produced by consuming small quantities of pulses. Bengal gram has the efficiency to reduce serum cholesterol level by increasing faecal excretion of total bile acids.

Lathyrism: Lathyrism is a paralytic disease seen in India and Spain where a type of pulse called khesari dal. — Lathyrus sativus — is used liberally in food. In India Khesari is extensively cultivated in Madhya Pradesh, Bihar and Orissa. In Uttar Pradesh, West Bengal, Maharashtra, Gujarat, Karnataka and Andhra Pradesh it is cultivated in limited extent. Kesari dal is very cheap and highly nutritious.

The flour prepared from this dal is used as a staple food in these areas. In these areas most of the agricultural labourers get khesari dal as their wage and about 250-500 gm of it is consumed by a person in a day. As a result about 4 per cent of population in these areas are crippled. Surveys in two districts of % Madhya Pradesh (Rewa and Satna) showed 32,000 paralytic victims with lathyrism. The cause of lathyrism is the production of a toxin known as BOAA-B oxalyl amino L alamine which affects the nervous system. The toxin is produced more by the busk and its consumption causes high incidence of disease. The Indian Council of Medical Research has developed certain methods to destroy the toxin.

The disease occurs in different stages. The onset is sudden. But in the first stage itself it can be detected. On exposure to cold the calf muscles contract and backache occurs. An awkward manner of walking with short jerky steps and crossed gait are common. Muscular stiffness necessitates the use of a stick to maintain balance.

Later spastic paralysis of the lower limbs, and exaggerated knee and ankle jerks occur. And at this stage the patient needs crutches for support. Finally correct posture becomes impossible as the knee joints cannot support the weight of the body. The patient crawls by throwing his weight on his hands. The serum globulin and urinary glutamic acid are increased in this condition.

The National Institute of Nutrition identified the toxic factor and a simple feasible solution was suggested by them to destroy the toxin. The household procedure is to steep the dehusked grain is boiling water. The seeds are drained of water and sundried and thus the toxin is eliminated. About 90-95 per cent of the toxin is leached out by this method.

Soyabean: Soyabean is otherwise known as the miracle bean because it is a storehouse of proteins, fat, vitamin A, thiamine, riboflavin and nicotinic acid. The protein content in soyabean is about 43 per cent and fat content is 19.5 per cent. The proteins of soyabean consists of all the essential amino acids in proper proportion except methinine and cystine.

It has a good amount of lysine and it can be used for supplementing rice which is deficient in lysine. Apart from the nutritive importance it is also made use of in the production of penicillin and other antibiotics for industrial purposes. Enamel paints, varnishes and printing inks are made out of soyabean.

Even though it is rich in nutrients, unless it is processed properly, proteins and carbohydrates of soyabeans are poorly digested. Soyabean contains a factor which inhibits the protein digesting enzyme, trypsin. Heat destroys it and after cooking if it is used the protein is properly digested. Soaking, steaming, germinating and fermenting are other processes which improve the nutritive value.

Soyamilk is very wantious, but soyabean seems to produce goitre at times because it interferes with iodine metabolism. Soyamilk has about 4.2 gm of 3.4 gm of fat and 0.75 gm of salts. Soyabean can be parched and used like parched bengal gram. It can be used as a substitute for Dal gram flour in recipes.

The Cooking: Soaking, use of sodium bicarbonate, acids, alkali or calcium salts quantity of cooked pulses. Dals absorb water before they can be cooked. If dal is soaked in cold water overnight or in warm water 1 4 to 5 hours it can be cooked within 30 to 40 minutes.

But if hard water is used for cooking it takes more time. (a) If sodium bicarbonate is added to cook pulses cooking time is reduced. (b) vitamin like thiamine destroyed. Addition of acid helps the cooking of pulses whereas as retard the cooking. Addition of formic acid of the pulse also influences the cooking quality. The exact causes for these variations are not found out.

Cereals: Cereals and millets form the staple food of human race. Wheat, rice, maize (corn), barley, oats, jowar, ragi and bajra are the common cereals and millets used. Cereals as a whole are rich sources of starch or carbohydrate, and good sources of protein in an Indian diet. Millets like ragi are a rich source of mineral calcium. Cereals are also good sources of B group vitamins.

Wheat: Wheat is one of our important food crops. Next to rice wheat is the most extensively grown foodgrain. Wheat varies in its nutritive value, especially its protein content depending upon the seeds, variety in soil and water supply. The protein content varies from 7 to 20 per cent. In the 'green revolution' new varieties of superior quality wheat were produced. Sonalika, safed lerma, kalyan sona, sharbati sonora are some of these new varieties of wheat.

The wheat grain consists of three main parts. The germ or embryo is rich in protein and oil, the endosperm is rich in starch and fair sources of proteins and the outer coverings such as pericarp. Various wheat products available are broken wheat, semolina, whole wheat flour (atta) and white flour. These flours are made from different kinds of wheat. Hard wheat, durum wheat and soft wheat are the different types of wheat from which flours are made. Flours of different percentages of extraction are obtained.

Depending on the extent of milling, flour made from hard wheat has an extraction of 83 per cent and it is used for chapati making. With 75 per cent extraction white flour or soft wheat products are obtained. Biscuits, cakes, crackers and breakfast items are prepared from this flour. Semolina is prepared from hard wheat and durum wheat. It is granular wheat and it is used for the manufacture of macaroni products like noodles.

Composition of Wheat Flour: Wheat is a very good source of carbohydrates and proteins; 100 gms. of wheat provides 350 calories. Wheat contains carbohydrate in the form of starch; 66 to 68 per cent of flour is carbohydrate. Wheat flour has cellulose and hemicellulose and its quantity enhances the quality of dough prepared from it

Protein: Wheat flour contains glutenin and glidin as proteins which are commonly known as gluten. Wheat flour contains about 12 to 12.5 gms of protein per 100 gm. The strength of wheat flour is based on die quality of gluten in it

Gluten is comparatively poor in amino acid lysine. Pulses, soya bean and groundnut are good in lysine content and they can be supplemented with wheat.

Lipids: Wheat flour contains 1.88 per cent fat; its kernel has about 3 per cent and bran contains about 4.8 per cent fat. Fat is present in the gluten of the flour.

Minerals: Wheat flour is poor in calcium content. The iron content of whole wheat flour is high but refined flour has less iron. Phosphorus content in whole flour is high compared to refined flour. But this high phosphorus content interferes with the absorption of calcium and iron.

Vitamins: Wheat is comparatively a good source of thiamine, riboflavin and nicotinic acid. But cooking results in loss of these B complex vitamins. During chapati making about 33 per cent of thiamine is destroyed and during fermentation much more thiamine is lost. In the preparation of puri 40 to 90 per cent of thiamine is lost. During bread making about 10 to 30 per cent lysine is destroyed. If the bread is toasted again this vitamin is lost Bulgar wheat is much more nutritious. Wheat is parboiled in the preparation of bulgar wheat. During this process vitamins and other nutrients in the outer layers of the seeds move to the centre. Thus bulgar wheat is better in its auditive value.

Processed Wheat Products: Bread, biscuits, toast and breakfast cereals like wheat flakes and shredded wheat are processed wheat products. The autrivtive value of bread, especially fresh bread, is similar to two-thirds of wheat flour in its quantity.

In the preparation of biscuits thiamine and protein is lost. On baking, apart of thiamine is lost and lysine is also lost. The average protein content in biscuit is 5 to 8 per cent as against 12 per cent in wheat flour. In the preparation of toast heat is applied on bread slices and part of thiamine is again destroyed. Lysine is also lost considerably. Readymade breakfast items are also inferior in nutritive value.

Rice: Rice is the staple food not only in India but also in many of the South-East Asian countries like Ceylon, Burma, China and Japan. Rice with its husk is called paddy. Husk is the loose covering of rice. It has no nutritive value. About 20 per cent by weight of paddy is busk. Removing the husk is called milling. Bran, aleurone layer, embryo or the germ and endosperm are the other layers. Bran is die outermost layer which contains nutrients and cellulose. Aleurone layer which lies beneath the bran is rich in vitamins, minerals, proteins and fats. Embryo or the germ is at the base of the grain which is rich in all nutrients and vitamin E. Endosperm forms 75 per cent of the grain and it comprises mainly starch. On milling vitamin B lost considerably. Parboiling improves the vitamin content of rice.

Parboiling of Paddy: In the conventional method of parboiling paddy is steeped in cold water for two to three days. Then it is steamed for about 5 to 10 minutes and men dried under the sun. During soaking paddy is fermented and at times gives out a foul smell. B complex vitamins in the pericarp and germ penetrate into the endosperm and get fixed into the grain. If parboiling is not done with care certain moulds present in the grain grow and produce toxins like aflatoxin. This can be rectified by using modern techniques in food processing. Through hot soaking process developed by CFTRI (Central Food Technological Research Institute) in Mysore is can be rectified. In this method paddy is soaked in water at 65° to 70°C for about 3 to 4 hours. Then water is drained and the soaked paddy is steamed in the same vessel for 5 to 10 minutes. This paddy is dried in sun or in mechanical driers. The nutritive value, especially of thiamine and niacin, is improved by parboiling.

Nutritive Value of Rice: Rice contains carbohydrate in the form of starch. Though it is a source of protein, it is a poor source of fat, iron and calcium, provides about 335 calories per 100 gm of dry weight. The protein content is about 7 per cent which is not an appreciable amount since it is consumed in large quantities, rice supplies a good amount to proteins.

In poor Indian diets cereals or millets are the main source of proteins. Though the protein content is low in rice compared to wheat the quality of rice proteins is superior to wheat proteins. Rice protein deficient in lysine which can be improved by supplementing the with pulses. High yielding varieties of rice are cultivated in our country. The protein count of hybrid variety rice like IR-20 is 9 per cent against 7 per cent of ordinary rice. Rice is poor in its mineral content especially calcium. Iron content is also negligible, therefore rice-eaters must depend on other sources of minerals. Green leafy vegetables can be used for miner supplementation.

Rice is a poor source of carotene or vitamin A, and vitamin B present in good amount. Since it is present mainly in the outer lay polishing reduces the vitamin content. As polishing is increased content of B vitamins decreases. In 20 per cent polishing thiamine (B. content is only 40 mg/gms whereas in parboiled rice at 20 per a polishing thiamine is 200 mg/100 gm. Niacin content is 0.6 mg/100 gm, in 20 per cent polished rice whereas in parboiled rice it is 2.8 mg/l gm.

Among people who depend entirely on polished rice, a burn sensation in the feet, tingling and numbness occur. This is due to deficiency of thiamine in the diet. Even though rice is a good source of B complex vitamins. Its high solubility in water makes it non-available after various processes of cooking. About 60 per cent of it is lost through washing, if rice is washed twice or thrice. In the cooking water , remaining B complex gets dissolved. If the "conjee" water is thrown i this amount is also lost. Cooking rice in just enough water and utilise the excess water, if any, are the useful tips to make use of this vitamin.

Gelatinisation: Starch is present in cereals. This is insoluble in water, but in hot water it dissolves and forms colloidal dispersions. It swells and forms a sticky substance. In cereals about 65 to 80 per cent starch is present. Next to cereals roots like tapioca, potato and sweet potato are with starch granules. These roots and tubers contain about 22 to 39 per cent of starch. In starch there are two components—amylose and amylopectin. Amylose is about 15 to 30 per cent and amylopectin about 70 to 85 per cent.

In different sources of starch it is present in different proportions. Amylose is soluble in water and amylopectin is insoluble in water. On 5 per cent concentration starch forms gels. Among the various sources of starch corn starch is the finest. This property of starch to gel is exploited in cookery. In many preparations starch is used as a thickening agent. To get a good gel, gelatinisation temperature of different starches must be known. Usually it starts to gel at 65° to 72°C. The enzymes in the starch are the other factors which affect the gelatinisation. Commercial products like sago, custard powder, corn syrup, glucose or dextrose and dextrimaltose are prepared from tapicao starch or from corn starch.

Millets: The common millets used in our country are jowar, ragi and bajra. Though millets are comparatively cheap they are superior in nutritive value. Usually people prefer rice or wheat as their cereals. If these millets are used more many deficiency symptoms among various groups can be minimised as they are excellent sources of vitamins and minerals.

Ragi: In India ragi is used mainly in Karnataka and Tamil Nadu as a staple food. In Kerala, ragi is used as a weaning food. Actually ragi is considered as a 'poor man's milk' by nutritionists because of its high calcium content. It can be easily substituted as a breakfast item. If we consider its cost it is cheap and at the same time rich in minerals that are lacking in rice or wheat. Like wheat or rice, ragi also contains 7 to 10 per cent protein and is equivalent to the above cereals in its carbohydrate content. It is considered superior because of high calcium content. Ragi is also a good source of iodine. No other cereal contains iodine.

Another special feature of ragi is that the carbohydrate in ragi is digested slowly and so it can be given to diabetic patients because the sudden rise of blood sugar can be avoided by this. In certain hybrid varieties of ragi protein and iron contents are more and thus it is superior in its qualities. Ragi can be incorporated with pulse instead of rice in all breakfast items.

Bajra: Bajra is used in certain parts of the country as a staple food. It consists about 12 per cent of protein and calcium and iron content is compared to rice. Bajra is a better source of vitamin B, especially thiamine and riboflavin.

Jowar: Jowar is used in Maharashtra, Andhra Pradesh, Karnataka, Tamil Nadu and Madhya Pradesh. The protein content in jowar is 10 per cent and the amino acid proportion is better. It has more B complex vitamins compared to rice. But certain varieties of jowar inhibit the availability; of niacin and cause the deficiency symptoms of pellagra.

Effect of varying concentration of sugar syrup and crystallisation

Boiling temperature° C	***Sugar percentage (Syrup)***	***Cold water test***
110-112° C	78-80	Does not form a ball
113-115° C	83	Soft ball, eg., fondant
118-120° C	86	Firmer ball, eg., caramels.
128-130° C	90	Hard ball, but still plastic, eg., candy
138-140° C	94	Very hard ball, eg, toffees
149-154° C	98	Very hard and brittle, eg., brittles.

To prepare the syrup add a small amount of sugar in water and heat, when the whole sugar is dissolved examine the consistency.

Minerals and Fats

In an adult 4 per cent of the body weight is from minerals. There are 24 minerals in the body. Mineral substances are important in maintaining good health because they are part of our tissues, skeletal structure and the body fluids. Hormones, enzymes and vitamins contain minerals.

Vital organs like heart and nervous tissues cannot function normally in its absence. Water balance, regulation of fluids, and the acid base balance in the body depend to a great extent on certain mineral balance in the body. Neuromuscular activities, coagulation of blood and fluid movement in the tissues are under the control of certain inorganic substances in the body.

Minerals are inorganic substance that are present in the body tissues like phosphomin, iron, sulphur, zinc or copper or soluble minerals in the body fluids like sodium chloride or mineral components in the bones and teeth and hormonal secretions. Water balance in the body is regulated by minerals like sodium and potassium. Our body has calcium, phosphorus, sodium, potassium, chlorine, iron, magnesium, copper, iodine, cobalt, zinc, aluminium, arsenic, bromine, fluorine, nickel, chromium, cadmium, selenium, silicon, vanadium, and molybdenum as minerals. Each one of this has some role in maintaining the structure and function of vital organs of our body. Few milligrams of a mineral element decide the skeletal growth or the mental ability of a person.

Optimum intake of minerals and favourable environment for its utilisation are equally important to maintain mineral balance. During the growth period and also during pregnancy, lactation and old age, certain minerals are very essential to maintain health and growth.

Calcium: The most abundant mineral element in the body is calcium. Nearly 99 per cent of the calcium in the body is present in the skeleton. The remaining part of calcium performs a variety of functions in the body. About 850-950 gms of calcium is present in a healthy man.

Functions: The most important function of calcium is the ossification of bones. If calcium content in the blood is below normal, bone calcium is depleted and serum level is brought to normal. Parathyroid hormone regulates calcium level in blood and its metabolism in the bone. The bones constitute a mineral reserve which may be drawn upon in need. Calcium is also essential for the formation of teeth and for the contraction of the heart and muscle.

The regulation of muscle irritability is dependent upon the serum calcium level. Calcium is also required for normal nerve impulse transmission and excitability of nerve fibres and nerve centres. Calcium forms part of acelytcholine which is necessary for the transmission of nerve impulses. Calcium helps in the coagulation of blood and is needed for permeability of capillary walls. Calcium exerts its influence in the absorption of vitamin B_{12} from the intestine. The mineral has been shown to activate certain enzymes, especially lipase and certain proteolytic enzymes, adenosine triphosphate and rennin.

Factors Affecting Calcium Absorption: The most important factor promoting calcium absorption is vitamin D content in the body. Utilisation of calcium is more when it is needed by the body. For example, during growth the absorption of calcium is more than in adults. Calcium content in the diet increases the absorption of calcium. Phosphates and calcium ratio in the diet also influences the absorption. If the ratio is 1:1, absorption rate is high. If phosphate ratio is high, calcium absorption is low. Phytic acid which is present in cereals combines with phosphorus and forms insoluble calcium salts and calcium phytate and inhibits calcium absorption.

Abnormal intake of fats and fatty acids reduces the calcium level in blood and its metabolism in the bone. The bones constitute a mineral reserve which may be drawn upon in need. Minerals like iron which forms insoluble phosphates interfere with phosphorus absorption which indirectly affects calcium absorption.

Absorption of calcium is facilitated by low intestinal pH. Normal acidity of stomach due to the presence of hydrochloric acid is essential to stimulate the absorption of calcium. Parathyroid hormone also aids in the absorption of calcium from the stomach. If cereals with husk are used in the diet, calcium absorption is reduced. Whole flours, oatmeal, and preparation with these, if not fermented, increases the inhibition of calcium utilisation. On fermentatión of the enzyme phytase splits up calcium phytate and then increases its absorption.

Like phytic acid, oxalic acid is present abundantly in leafy vegetables and certain fruits. This forms an insoluble compound calcium oxalate and thus diminishes the absorption of calcium. Magnesium, if taken in abundant quantities, often interferes with calcium absorption and results in poor calcium absorption. During sprue, steatorrohea or in biliary fistula fat is absorbed poorly and thus accumulated fats help the formation of calcium soaps. Good quality protein content in the diet helps calcium absorption. During rickets citric acid increases calcium absorption.

There are many factors which hinder the absorption of calcium. Phytic acid, oxalic acid, magnesium and fat reduce the absorption of calcium. Cereals contain phytic acid which combines with calcium to form insoluble calcium phytates. Phytic acid is present in the outer parts of certain proteolytic enzymes.

Deficiency: Calcium deficiency symptoms are usually accompanied by a lack of phosphorus and vitamin D. Any of the factors which affect calcium absorption can lead to its ineffective utilisation. Vitamin D deficiency alone can produce calcium deficiency symptoms. Pure calcium deficiency is osteoperosis. In osteoperosis serum calcium level is normal but body store is reduced.

During old age the capacity of the body to absorb and utilise calcium is diminished. In women hormonal imbalance during menopause or removal of ovaries at earlier stages results in osteoperosis. Corticosteroid injection for any purpose may result in osteoperosis. In osteoperosis pores are seen in the bones due to decalcification. Back pain, loss of weight, fracture of bones for no reason and loss of teeth are some of the clinical symptoms. Calcium and vitamin D deficiency results in rickets and osteomalacia among children and adults, respectively.

Tetany is another symptom of calcium deficiency. In tetany there is hyperexcitability of the nervous system. A series of spasmodic movements take place in the muscles. Depending upon the level of serum calcium symptoms of tetany vary. Irritation, confusion, pain, involuntary muscle spasm, bronchial spasm, spasm in the gastro-intestinal tract, pain and convulsions are again symptoms of tetany.

If parathyroid gland is in hypoactivity, calcium is not properly assimilated and results in chronic tetany. In this condition nervous changes occur and impairment of memory takes place. Depression, irritation, loss of hair and nails, roughness of the skin, pitting of teeth and epileptic fits are other symptoms of this condition. Cataract is common in such patients.

Hypercalcemia: Hyper-vitaminosis-D or Hyperparathyroidism produces hypercalcemia. Normally infants of 5-8 months get this problem. Loss of appetite, vomiting, constipation and flabby muscles are the symptoms.

Sources of Calcium: Milk and milk products are the best sources of calcium. Calcium is absorbed maximum from milk and milk products because of the presence of lysine and lactose in it. Ragi is the cheapest source of calcium; therefore it is known as 'poor man's milk'. Sesame seeds and leafy vegetables are also good sources of calcium. But the availability from these sources is not insured as in sesame seeds calcium is present in its husk and on dehusking it is removed. In leafy vegetables the presence of oxalates inhibits its absorption. Molasses and vegetables like peas, beans, potatoes, pulses, and small fish and dried fruits are rich in calcium. Unabsorbed calcium is excreted through faecal matters.

Requirements: The FAO/WHO expert group (1984) has recommended 400-500 mg of calcium per day for an adult man and woman. During pregnancy and lactation the requirement increases to 1,000 mg/day. For infants 500-600 mg and for children of between 1 and 9 years 400 to 500 mgs and for age group 10 to 15 years 600-700 mg are recommended.

Phosphorus: No other mineral takes part in a number of functions as phosphorus in different parts of our body. An adult human body contains about 400-700 mgs of phosphorus. About eighty per cent of the body's phosphorus is deposited in the bones and teeth as calcium phosphate. The remaining portion is located as sodium phosphate and potassium phosphate in the body fluids and in cells. Phosphorus is also present in many enzymes and co-enzymes and co-carboxylase. Thus it takes part in metabolism.

Functions: Phosphorus takes an important role in the calcification of bones. It also plays an essential part in carbohydrate metabolism in phosphorylation of glycogen. Phosphorus is an essential constituent of co-enzyme and co-carboxylase enzyme system in the oxidation of carbohydrate, proteins and fat metabolism.

Phosphorus compounds control the storage and release of energy through ATP, ADP system. Phosphorus is present in the nucleoproteins which are an integral part of the cells. Phospholipids, which are present in various cells of our body, contain phosphorus. Inorganic phosphates in the body fluids act as a buffer system in the body.

About seventy per cent of ingested phosphorus is absorbed. If iron, aluminium and magnesium are present in excess amount in the diet they interfere with the absorption of phosphorus. The proportion of calcium and phosphorus is an important factor which regulates its absorption.

Requirements: The daily requirement is about 1 gm. Phosphorus requirement is similar to calcium for various groups.

Sources: Animal foods like meat, fish, poultry and eggs are excellent sources of phosphorus. Milk and milk products are good source of phosphorus. Though cereals and legumes are good sources, phosphorus is less available from them as it is present in the form of phytic acid which is not utilised well.

Iron: Iron is a component of the complex protein haemoglobin, a haemoprotein component of the red blood cells. It is the pigment of red blood cells. Of the 3 to 5 gms of iron present in our body more than half of it is present in haemoglobin. The rest of it is stored in the liver, spleen, bone marrow and muscles. In the muscle it is present in the form of myoglobin. Iron is also present in some enzymes like cytochromes, catalase and peroxidase.

Functions: The most important function of iron is to form a constituent of haemoglobin which takes part in the transportation of oxygen from the lungs to the tissues. This oxygen is used by the body for the processes of metabolism. In its deficiency the oxygen-carrying capacity of the blood is reduced. In red muscles myoglobin, a pigment substance, is formed with the participation of iron.

This also has the same oxygen-transporting function like the haemoglobin. During strenuous activity this muscle is involved. In our body there are certain cytochrome protein pigments with iron as an essential component. They act as respiratory enzymes. Certain flavin enzymes also contain iron in them.

Absorption of Iron: Dietary iron is absorbed primarily from the duodenum by the mucosa. Depending upon the need of iron in the body its absorption varies. If the haemoglobin concentration of the blood is satisfactory, little or no iron is absorbed and it is excreted in the faeces. Ferrous iron is absorbed more efficiently than ferric iron. In the intestine most of the iron is reduced to ferrous state. Absorption of iron is enhanced in an acid medium.

Vitamin C also enhances the absorption of iron, in the intestinal tract, because of its strong reducing capacity. Protein content in the diet helps the absorption of iron to some extent Sulphur containing amino acids reduce iron which helps its absorption. Excess of calcium phosphate, phytates and oxalic acid interfere with iron absorption by forming insoluble iron salts or oxalates. Minute traces of copper help the formation of haemoglobin. The rate of absorption of iron is affected by the magnitude of iron store in the body and the rate of erythropoiesis. Malabsorption syndromes also reduce iron absorption.

Deficiency of Iron: Anaemia is a condition where the haemoglobin level is lowered in the blood. Shortage of iron in the diet is the most common cause of anaemia. Iron, folic acid, proteins, B_{12}, copper, cobalt, vitamin C and pyridoxine are the nutrients required for erythropoiesis. Folic acid deficiency produces megaloblastic anaemia.

Symptoms of Anaemia: When the haemoglobin level falls below 12.5 gm/100 ml of blood, it results in the diminished oxygen-carrying power of the blood. Iron deficiency anaemia is otherwise known as hypochromic anaemia. In this condition the number of RBC is not much reduced but the quantity of haemoglobin is less. The pallor of skin and tissues, fatigue, weakness, giddiness, blurred vision, anorexia, heartburn, palpitation, dysproea, oedema over feet, headache and a feeling of tiredness are some of the symptoms. The nails show koilonychia and tongue and papillary atrophy are seen. Among women menstrual disturbances like menorrhagia or phases of amenorrhoea take place.

Nutritional megaloblastic anaemia occurs due to dietary deficiency during old age, increased requirement during pregnancy, lactation, prematurity, rapid growth and in endocrine disturbances. Haemolytic anaemias occur during malignancy, rheumatoid arthritis, pulmonary tuberculosis and in skin diseases. Alcoholism and side effects of drugs also result in anaemia. In diseases like idiopathic steatorrhoea, tropical sprue and in partial gastractomy anaemia takes place. During pregnancy this type of anaemia usually occurs in the second trimester. Weakness, palpitation and fatigue are followed by fainting bouts.

Anaemia must be treated with ferrous sulphate tablets and iron-rich diet.

Hemosiderosis: It is a disorder of iron metabolism. In this condition large deposits of iron are made in the liver and in the recticulo endothelial system. Iron with carrier protein which is known as transferrin of circulation becomes saturated with iron. When transferrin is unable to bind all the absorbed iron this metabolic disorder of iron takes place. If iron vessels are used for almost all cooking; much more iron is ingested in the body, which may result in hemosiderosis. The Bantu population of Africa is an example of this condition.

In hemolytic anaemia a large number of red blood cells are destroyed and this can lead to hemosiderosis. Prolonged prophylaxis with massive doses of iron can often produce hemosiderosis.

Sources of Iron: Liver, kidney, heart, lean meat, egg yolk and shellfish are the best sources of iron. Dried beans, legumes, dried fruits, nuts, green leafy vegetables, whole cereals, enriched grains and molasses are good sources. Milk is a poor source of iron.

Requirement of Iron: Recommended intake of iron is calculated, based on its loss and absorption. For an adult man weighing 55 kg the iron requirement is 24 mg/day and for a woman of 45 kg weight 32 mg is the requirement. During pregnancy 40 mg and during lactation 32 mg of iron is required.

For infants till one year 1 mg/kg body weight and for children of 1 to 9 years of age 20 to 25 mg of iron are required. For boys and girls between 10 and 12 years of age 30 to 25 mg and for boys 13 to 15 years of age 25 mg and for girls from 13 to 15 years 35 mg of iron are required. During later adolescents for boys (16-18 years) 25 mg and for girls 35 mg is the recommended allowance.

Magnesium: The human body contains minute quantities of magnesium in the bones along with calcium and phosphorus and the rest in soft tissue. An adult human body has about 25 gm of magnesium.

Functions: Magnesium plays an important role in normal calcium and phosphorus metabolism in man. It is also present in certain enzymes like co-carboxylase which decarboxylase pyruvic acid. Magnesium also takes part in the phosphorylation in metabolic pathways.

Magnesium is easily absorbed from the intestine. Excess magnesium is excreted in the faeces. Normal human serum contains about 2 to 3 mg/100 ml.

Deficiency: Prolonged starvation brings magnesium deficiency, vomiting and intestinal fistula. Diabetic acidosis is associated with raised serum level of magnesium. Neuromuscular irritability, tremors, cramps, twitching and convulsions occur in its deficiency. Confusion, disorientation and visual hallucinations are some other complaints with magnesium deficiency.

Sodium: Sodium is essential for the normal functioning of the body. Sodium plays an important role in the regulation of acid-base balance and water metabolism in the body. Sodium in the form of sodium chloride is ingested directly through food and many food materials contain this mineral. Almost all the body sodium is found in the extracellular fluids of our body. Through sodium the osmotic pressure is also regulated by the body. A healthy kidney regulates the excretion of sodium according to the requirement.

During kidney diseases sodium is retained in the body and oedema takes place. Under normal conditions 90 per cent of ingested sodium is excreted in the urine in the form of sodium chloride and sodium phosphate. Sweat contains some amount of sodium. Sodium helps to contribute alkalibility of the gastrointestinal tract.

Sodium plays a special role in controlling the heartbeat by helping in its origin and maintenance. Contraction of muscles and regulating the permeability of cell membranes are the other functions of sodium.

The adult human body contains about 100 gm of sodium. On an average diet about 3 to 5 gins of sodium is excreted in the urine. In profuse perspiration sodium chloride is depleted from the body. In sodium depletion muscular cramps, weakness, headache and vascular collapse take place. In decreased water intake the serum level of sodium is increased and this condition is known as hypernatraemia. Blood urea is increased and cellular water is depleted during dehydration and in hyponatraemia.

In infants shrinkage of brain causes rupture of cerebral veins and further cerebral thrombosis. Retention of sodium produces oedema and ascites. In renal diseases, cardiac failure, hepatic cirrhosis with ascites, toxaemia of pregnancy, prolonged administration of corticosteroids sodium must be restricted through diet. Normally 4-5 gm sodium is injested through food and it is restricted to 250 mg, 500 mg, or 750 mg depending upon the condition.

Requirements: For an adult 1-2 gms of sodium and for children 350 mg-1 gm is recommended. For infants 150-400 mg is enough.

Sources: Animal foods like meat, meat extractives, fish, poultry, milk, milk products, nuts, bakery items, salted biscuits, egg-white, dried fruits, roots like beetroot, carrot, radish and leafy vegetables like spinach are excellent sources of sodium.

Abnormalities of Sodium Level in Blood: Hyponatraemia and hypernatremia are two conditions which manifest the sodium level in the blood.

Hyponatraemia: When serum sodium level falls below normal the condition is known as hyponatraemia; various conditions lead to this symptom. Severe dehydration, decrease in blood volume, decrease in blood pressure, circulatory failure are some of them, prolonged vomiting, diarrhoea, chronic renal diseases with acidosis, Addison's disease due to adrenocestical deficiency are the common causes for hyponatraemia.

Hypernatraemia: In this condition the plasma sodium iron content is higher than normal. Hyperactivity of adrenal cortex as in cushing's syndrome, prolonged treatment with cortisone, ACTH and sex hormones are some of the causes of hypernatraemia. Deficient water intake due to vomiting, nausea, difficulty in swallowing with sore throat, oesophageal disease, cerebral impairment, increased water loss through respiration diarrhoea, excessive urination in diabetes mellitus, chronic renal diseases, increased sodium intake and deficient sodium excretion brings hypernatraemia.

Potassium: In contrast to sodium, potassium is present in the intracellular fluids of the body. It is an important constituent of cells. In adult body 250 gm of potassium is present. Most of it, that is, 97 per cent is found in the intracellular fluids and the remaining 3 per cent in extracellular fluids. The other important function of potassium is that it influences the cellular excitability of smooth, skeletal, cardiac and nervous tissues. If potassium content is high muscular irritability is increased. Diminished calcium and magnesium also result in such a condition. Involuntary muscular function is controlled by potassium content in the body. When the potassium level in the myocardium is changed it is reflected in the electrocardiogram. Potassium is also necessary for the acid base balance in the body.

Potassium deficiency shows weakness in muscular functions, mental disorientation and cardiac irregularities. Hyperpotassemia results from kidney diseases when renal excretion is less. In dehydration these symptoms occur. Deficiency of potassium is not common unless as a complication of diseases. Deficiency condition shows hypokalaemia during operation when food cannot be ingested. In diabetic acidosis this condition occurs. Deficient intake due to starvation and excessive loss from gastro-intestinal tract leads to deficiency.

Hypokalaemia occurs due to deficient intake of it while a surgical patient is exposed to prolonged parent therapy, excessive loss from gastro-intestional tract through persistent vomiting, fistulae or diarrhoea before and after surgery, the patient may be on a poor diet which may enhance the excretion of potassium. Surgical cases due to low intake produces hypokalaemia.

In diabetic acidosis potassium is lost through dehydration on insulin therapy, sugar is converted to glycogen and starch. This glycogen formation depletes some potassium from extracellular fluids and serum potassium level falls and it is dangerous.

While diuretics are administered considerable loss of potassium through urine leads to hypokalaemia. Nephritis and other kidney diseases, tumours of large intestine, treatment with steroids, deficiency disease like kwashiorkor, megaloblastic anaemia, heatstroke and violent exercise are some of the other causes which produce hypokalaemia. Prolonged hypokalaemia causes injury to myocardium and kidney.

Hyperkalaemia: When the potassium level in the serum is high the condition is known as hyperkalaemia. This occurs due to diminished urine volume, intake of potassium salts of penicillin, iodide, acidosis, endogenous tissue breakdown, Addison's disease, kidney failure due to kidney disease or chronic renal failure and rapid intravenous administration of potassium are some of the reasons for hyperkalaemia. Cardiac and the central nervous system are affected by this condition. Anorexia, weakness of the muscles, diminished tendon jerks, low heart sounds followed by peripheral vascular collapse and general apathy are some of the symptoms.

Trace Elements: Elements which occur in our body in small quantities are known as trace elements. Copper, iodine, zinc, manganese, molybdenum, fluorine, selenium, nickel, chromium, cadmium, silicon, vanadium, strontium, cobalt and chlorine are some of the trace elements. Though these trace elements are present in traces they are very essential to maintain the structure of various organs in good condition. RBC formation, thyroxine production, formation of healthy bones, haemoglobin formation and proper enzymatic actions are some of the important functions of trace elements in our body.

Iodine: The significance of iodine as an essential trace element lies in its role in thyroxine production. Thyroxine is the hormone produced by the thyroid gland. Iodine is the/ main component of thyroxine and a normal -adult has only about 10 mg of iodine in the thyroid. The absorbed iodine combines with a thyroxine binding protein. Thyroxine plays an important role in carbohydrate metabolism. Deficiency of thyroxine interferes with the normal physical and skeletal development of the child during the growing period. Iodine also helps the tissues in its oxygen consumption. Conversion of glycogen to glucose occurs only in its presence. Thyroxine increases the heart rate and depletes bone calcium.

Deficiency: If the iodine intake is deficient the thyroid gland enlarges and this condition is known as simple or endemic goitre. The normal thyroid gland weighs about 25 gin. In goitre it enlarges from 200 to 500 gm. Iodine deficiency goitre results in hypofunctioning of the gland. If diagnosis is done early and iodine supplements are given the thyroid may become normal. If timely treatment is not given the enlargement will become permanent. During the growing period if severe deficiency occurs it results in retarded growth and the condition is known as cretinism.

People living mainly in submountain zones are affected. In India residents of sub-Himalayan regions and Kashmir and parts of adjoining plains are the victims. The most affected are among girls between the ages of 12 and 18 years and boys from 9 to 13 years. The areas where goitre is common is known as the goitre belt. According to the latest data available, about 40 million people are affected with goitre. In these areas air, water, soil and foods grown are poor in iodine which results in its high incidence. Goitre can be prevented by the regular use of iodized salt.

Goitrogenic Substances in Foods: Certain foods of brassica genus like cabbage, cauliflower, turnip, soyaflour and oil contain substances which react with iodine present in the food and make it unavailable. These substances are known as goitrogenic substances. Consumption of these foods in large quantities makes the iodine non-available to the body.

Requirement of Iodine: An adult requires about 0.15 to 0.2 mg of iodine and for infants and children 0.05 to 0.1 mg daily is enough.

Sources: Iodine is present in minute quantities in all foods.

The iodine content of foods depends on the iodine content of the soil. In certain parts the soil contains less iodine and the foods grown in such places are iodine poor. The soil of mountainous regions is usually iodine-poor soil. The best sources of iodine are seafoods, common salt and vegetables; cereals and legumes are poor sources of iodine. Most of the iodine in the vegetables is in their leaves and flowers.

Copper: Copper is found in all tissues of our body. It is an essential element in the formation of haemoglobin. About 100 to 150 mg of copper is present in the human body. The brain, liver, heart and kidney contain the highest amounts. In the blood, copper is present in the plasma and RBC. Copper is absorbed rapidly into the blood.

Copper is a component to certain enzymes in our body. It is part of the enzyme cytochrome oxidase which is essential for glucose metabolism. It is also part of tyrosinase, an enzyme necessary for converting the amino acid tyrosine to melanin, the dark pigment of the skin. Copper stimulates the absorption of iron. Copper is a constituent of the elastic connective tissue protein elastin.

Deficiency: Copper deficiency is not common in human beings. In malnourished persons iron is not absorbed properly and anaemia occurs. Premature babies fed on cow's milk develop anaemia due to copper and iron deficiency.

Requirement: An adult person requires about 2 mg/day. During pregnancy and lactation the copper requirement is increased to 3 mg/day. Infants from 0 to 12 months require 0.5 mg to 1 gm and children 2 gm and adolescents 3 mg of copper.

Sources: Millets like cholam, bajra, raw rice, whole grams, cow gram, dried peas, red gram dal, vegetables like drumstick are good sources of copper. Liver is the richest source of copper. Organ meat like kidney is also a good source of copper.

Zinc: In an adult 2 to 3 gms of zinc is present. It is present in certain parts of the eye, certain glands and in their secretion. Liver, muscles, bone and hair contain some of the remaining zinc. Blood, especially the RBC, has some zinc. In small amounts it is present in tissues and bone cells.

The important function of zinc is its role in enzymatic action. At least in 25 enzymes of digestive and metabolic action zinc takes part. It is also present in an enzyme carbonic anhydrase which is essential for the transport and elimination of carbon dioxide. Co-carboxypeptidase and alkaline phosphatase contain zinc in them. Zinc plays an essential role in the formation of DNA and RNA. It is a constituent of insulin and it is necessary for the formation of connective tissues. Zinc also aids in the healing of burns and wounds.

Deficiency: Zinc deficiency in animals results in growth failure, loss of appetite, degeneration of the male sex glands, testes, skin lesions, alopaecia or loss of hair, changes in the epithelial and cutaneous tissues, dermatitis and scaling and cracking of the paws. Abnormal bone formation, thickening and metabolism also occur during zinc deficiency. In human beings zinc deficiency produces anaemia, growth retardation and delayed genital maturation.

Requirement: Infants require 3 to 5 mg. For children 10 to 15 mg and during pregnancy and lactation 20 to 25 mg of zinc are required.

Sources: Seafoods, meat poultry and eggs are good sources of zinc. Cereals, legumes, and nuts contain considerable amounts. Fruits and vegetables are poor sources. On milling eighty per cent of zinc is lost

Cobalt: In an adult body about 1.1 mg of cobalt is present. It is found in the blood, liver, kidney, spleen and pancreas. It is also present in minute quantities in all tissues.

The most important function of cobalt is that it forms an integral part of vitamin B_{12}. In animals, if the feed is deficient in cobalt severe anaemia occurs. Addition of cobalt to feed or injection of vitamin B_{12} cures this condition. In human beings vitamin B_{12} synthesis occurs in the colon. At that phase absorption is not taking place. So preformed B_{12} must be ingested for the cobalt requirement. But deficiency of cobalt is not common among human beings. Cobalt also takes part in the enzymatic action and also in thyroid function.

Requirement: Since deficiency is not known, the requirement is not studied. In pernicious anaemia B_{12} deficiency is present and on administration of B_{12} it disappears.

Sources: Cereals, beans, peas and organ meats like kidney and liver are good sources.

Manganese: About 20 mg of manganese is present in an adult. Bones, liver, pancreas, kidney and pituitary gland contain manganese. The manganese functions in many enzyme systems. It activates certain enzymes which take part in the digestion and metabolism of carbohydrates, proteins and lipids. In the synthesis of ATP cholesterol, fatty acids and RNA, the manganese plays an important role.

Manganese deficiency in human beings is not known. Experimental deficiency symptoms are slowing of growth, abnormalities in bone, depressed reproductive functions, neurological weakness of the limbs, ataxia and lack of balance. Among birds a condition known as perosis occurs on manganese deficiency. It is characterised by bone deformities and dislocation of bones.

Sources: Cereals, bran, dried beans, peas, green vegetables and nuts are good sources. Tea and coffee are high in manganese content. But animal foods are relatively poor in it.

Chromium: An adult body contains about 6 mg of chromium. Chromium is present in skin, hair, muscles, brain, tissues, adrenal glands and in body fats.

An important function of chromium is its role in glucose tolerance factor formation in the body. Chromium niacin complex factor helps the action of insulin on glucose. Chromium is also essential for the activating enzymes involved in carbohydrate, protein and fat digestion and metabolism. The serum cholesterol level was raised on experimental animals in chromium deficiency.

Requirement: The exact requirements is not known. A normal balanced diet supplies enough chromium.

Sources: Organ meats like liver, meats, wholegrain cereals and brewer's yeast are good sources of chromium.

Fluorine: Fluorine is present in the body as fluoride. It is present in teeth, bones, thyroid gland, skin and liver.

Fluoride is necessary for the formation of caries-resistant teeth. Small amounts of fluorine prevent dental cariesin children. It promotes calcification.

Fluoride and Tooth Structure: For healthy tooth formation some amount of fluorine is essential. Though the exact mechanism by which it prevents dental caries is not fully understood, it is assumed that fluorine is incorporated into the tooth enamel. This reduces the solubility of enamel in the acids produced by bacteria. In its absence the enamel is soluble in the acid and tooth decays. In experimental animals fluoride seems to help normal growth during the growing period, but in human beings its role is not yet proved.

Toxicity or Fluorosis: If water contains more fluorine, dental fluorosis occurs. The enamel of the teeth loses its natural lustre and becomes rough. Opaque white Patches with yellow or brown stains are found irregularly on the teeth. Ridges or diffused spots and pitting occur over the teeth. The surface of the teeth gives a corroded appearance.

Skeletal Fluorosis: The bones of vertebral column, especially the cervical part of the spine, Pelvis and bones of the lower extremities, are affected in this condition. Hypercalcification of the bones and the collagen take place due to excessive fluorine deposition. Stiffness and pain are the earlier tymptoms. Weakness and spasticity of lower limbs, wasting of the distal muscles of hands, deep jerks on lower extremities and numbness or girdle pain are the other symptoms of skeletal fluorosis. Gradually the person becomes crippled as he cannot bend or squat for become stiff.

Requirements: Water containing 1 to 2 ppm prevents caries and so it is considered as the requirement.

Molybdenum: The adult human body contains only 9 mg of molybdenum. It is present in the liver, kidney, adrenal glands, bones, skin and tissues. Molybdenum is essential for the normal enzymatic system and its, function. Enzymes like xanthine oxidase and aldehyde oxidase make use of molybdenum. These enzymes are essential for the proper nucleic acid and iron metabolism. Molybdenum seems to aid in preventing teeth decay. The requirement of molybdenum is not known. Whole-grain cereals, dried beans, peas and dark green leafy vegetables are good sources. Meat is also an excellent source.

Fats and Oils

Fats and oils are used in cookery to enhance flavour and taste of food. Moreover, fat gives much more energy to body; from every gramme of fat 9 calories are given. In India most of our fats and oils are extracted from vegetable oilseeds. A solid fat-like ghee has a decisive place in Indian cuisine and it is often substituted by vanaspati. Liquid vegetable oils are hydrogenated and converted to solid fats. Hydrogenated fat is incorporated with vitamins A and D.

The Oils: Common oils used in cooking are groundnut oil, sesame oil, mustard oil, safflower oil, coconut oil, soyabean oil, cottonseed oil, palm oil and sunflower oil. Choice of fat for cooking depends on various factors. Flavour, taste, saturation of fatty acids and presence of double bonds, melting points, smoking temperature and nutritive value are some of them.

Liquid oils rich in linoleic acid like safflower, cottonseed, and groundnut are good in suppressing the formation of scales on the walls of blood vessel. The characterisitc flavours of various oils are due to the presence of very small quantities of certain substances which have no nutritional qualities. Cooking in fat gives fuliness of satiety along with taste to the dishes. Now most of the oils are refined to make them colourless and odourless.

Composition of Fats: Fats and oils are classified into two groups based on their origin as vegetable and animal sources.

Vegetable fats and oils are again broadly divided into three groups based on their composition. They are: Group I which has fatty acids with 16 or 18 carbon atoms; Group II has erucic acid; and Group III has short chain fatty acids.

Groundnut oil, cottonseed oil, sesame oil, sunflower seed oil, sunflower seed oil and soyabean oil are with 16 or 18 carbon atoms and they are termed as Group I. Mustard oil and rapeseed oil have erucic acid in it and they are in Group II. Coconut oil and palm oil are with short chain fatty acids in Group III.

Oil	***Linoleic acid content (%)***
Coconut oil	2
Mustard oil	15
Ghee	2
Palmoil	10
Palmolein	12
Safflower oil	70
Cottonseed oil	50
Sunflower oil	50
Groundnut oil	40

Animal fats are also classified into three groups based on their composition. Group I is milk fat which contains butyric acid and short chain fatty acids. Group II comprises animal body fat with fatty acids of 16 to 18 carbon atoms. Ox fat, pig fat and egg fat are examples of this group. In Group III fish oils with large amounts of polyunsaturated fatty acids are present herring, cod, halibut and sardine oils fall in this group.

Melting points of fats depend on the saturation of fatty acids. Solid fats like lard with high saturated fatty acids have high melting point and liquid oils have a low melting point.

Smoking Temperature of Fats: When fats and oils are heated to a high temperature decomposition of fats occurs and at a point visible fumes are given off. This temperature is called the smoking temperature of fats. The smoking temperature of fat is low when amounts of free fatty acids are high.

Smoking Temperature of Some Fats and Oils

Oil or fat		*Smoking temperature*
Butter	...	208° C.
Cottonseed oil	...	322° C.
Coconut oil	...	138° C.
Lard	...	194° C.
Olive oil refined	...	234° C.
Groundnut oil	...	162° C.

Fat absorption is high if foods are cooked in fat with low smoking points.

Smoking temperature of fat varies depending on the surface area of oil exposed while heating. The penetrability of the produce to be fried also determines the fat absorption while frying.

Shortenings: Fat is used as a shortening agent due to its ability to give a soft crisp texture to fried or baked products. In bakery items fat plays a significant role.

In wheat products fat brings shortening of gluten strands and in other dishes shortening is through gelatinisation of starch. To bring shortening effect one fat or a mixture of fats is used. In biscuit-making plastic shortenings are used which contribute a white, opaque and creamy appearance to the shortening. Butter and ghee are generally preferred as shortening agents. In certain preparations fat blends homogeneously with non-fat substance to form an emulsion and examples of such preparations are sweets like halwa, mysore pak or cake.

Changes in Fats: Changes in fat during heating are related to its smoking points. Free fatty acid is increased on heating. This is due to the change brought about by the moisture content of food materials. It hydrolyses the cooking fat and increases the free fatty acid content. Therefore, deep frying foods in fats with high smoking temperature is preferred. Ghee has a lower smoking point and therefore other oils or hydrogenated fat are suggested for deep frying.

When fat is used for deep frying, temperature control is essential. If high temperature is used to heat fat, it affects its nutritive value and in experimental cases it produced a toxic effect. Certain changes take place in fat during heating and this brings down the unsaturated fatty acid content. Repeated heating of fats brings chemical changes and results in poor absorption.

It also increased blood cholesterol level in experimental animals. If the fat is heated too high, the food will char outside and remain raw inside. If the temperature of fat is too low the food will get soaked in it. In correct temperature food will be crisp and uniformly cooked. To get the best result fat must be heated to its smoking point. When fumes appear insert food material, fry and lower.

Fats and Oils: The heat after frying the first batch allow the fat to heat to required temperature and add the second batch to fry. Absorption of fats by food materials varies depending on its smoking temperature. If fat with low smoking point is used to fry, more fat is absorbed and fat with high smoking point is required less to fry items. Food items with milk, sugar, or eggs require low temperature to fry them. If food items are fried at high temperature more fat is absorbed. If the food item has greater surface area fat absorbed is more and if surface area is small less fat is absorbed. The moisture content of food item also affects fat absorption. With more moisture content fat absorption also increases. Hard materials also take more fats.

Characteristics of Some Common Fats

Fat source fatty acid	*Poly unsaturated fatty acid*	*Saturated fatty acid*	*Mono-unsaturated*
Groundnut	31	19	50
Sesame	43	14	43
Safflower seed	67	7	26
Cottonseed	47	23	30
Coconut oil	2	92	6
Mustard oil	63	5	32
Butter	3	70	27
Soyabean	64	18	18
Lard	10	96	40

Salad Dressing: Fat is used as an emulsion in salad dressing. Along with fat, egg yolk, vinegar, salt, spices and starch are used in salad dressing. Mayonnaise, french salads and cooked salad dressing are the common methods used. Winterised oils are used in all these methods.

Mayonnaise: It is a semi-solid emulsion of vegetable oils, egg yolk or whole egg, vinegar and seasoning. Egg yolk functions as a seasoning agent.

French Dressing: Vinegar, oils (soyabeans, cornoil, sesame oil or sunflower oil) and salt and spices are added. It is mixed vigorously. Mustard powder is also added.

Lipids: Lipids are heterogeneous groups of compounds with same properties. Fat is the common household name given to lipids.

Like carbohydrates, fats are organic compounds composed of carbon, hydrogen and oxygen. But they differ from carbohydrates in that they contain much less oxygen and much greater proportions of carbon and hydrogen. Seventy-six per cent carbon, twelve per cent oxygen and twelve per cent hydrogen are present in fats. True fat has one molecule of organic ester of glycerol and three molecules of fatty acids. Fats vary in their structure due to the presence of different fatty acids with glycerol molecules. If the glycerol in a fatty acid is combined with three fatty acid units such fats are called triglycerides. Fats are insoluble in water and soluble in organic solvents like ether, benzene or chloroform. Their cooking properties depend upon the type of fatty acids present in them.

Classification: Lipids are classified into (a) simple lipids, (b) compound lipids, (c) waxes, and (d) derived lipids.

Simple Lipids: These are esters of fatty acids with various alcohols. Fats are esters of fatty acids with glycerol Oils are liquid at 20°C while fats are solid at 20°C. Combination of fatty acids and sterol alcohol form cholesterol. Food fats contain neutral fats. Fat is stored in the body as adipose tissue. This is also neutral fat.

Compound Lipids: When fatty acids and glycerol combine with other organic compounds they are known as compound lipids. Examples of compound lipids are phospholipids and glycolipids.

(i) *Phospholipids* (phosphatides) are substituted fats in which phosphoric acid and a nitrogenous base take the place of one fatty acid radical. These are present in vegetable oils and in the fats of animal tissues in the blood and in egg yolk. Phospholipids help the absorption of fat in the body, and also help the transfer of fat from one place to another. They have prominent role in the functions of the nervous tissue. Lecithin and cephalin are the two important forms of phospholipids in our body.

Lecithin is composed of glycerol, two fatty acids in ester linkage, phosphoric acid and choline. This is the most common form of phospholipid.

Cephalin contains nitrogen compounds other than choline, phosphoric acid, glycerol and fatty acid esters. Phosphatidyl ethanolamine and phosphatidylserine are the common forms of cephalin.

(ii) *Glycolipids* are compounds which contain a carbohydrate molecule with fatty acids and a nitrogenous base. Cerebrosides are the best example of glycolipids. Cerebrosides occur largely in the brain. Galactolipids, glycolipids, and sulfolipids are examples of cerebrosides.

Waxes: These are esters of higher fatty acids and alcohol other than glycerol. Cholesterol esters, sterol esters, esters of fat-soluble vitamins are the common waxes.

Derived Lipids: Substances derived from one fat on hydrolysis or by enzyme activities are called derived fats. Examples of derived lipids are fatty acids like stearic acid, linoleic acid, butyric acid, alcohols like glycerol, cetanol and lanol. Sterol alcohols like cholesterol, ergosterol and sistosterol are also derived fats. Sterols are nutritionally important. They are classified as animal sterols, plant sterols and mycosterols.

Cholesterol is the precursor of steroid hormones and bile salts. Hormones like progesterone, oestrogen and adrenal corticoids contain cholesterol. Sodium glycocholate and sodium taurocholate and the bile salts have cholesterol in them. Cholesterol esters produced by the liver regulate the cholesterol content in the blood. It is present in all animal and human tissues. The white matter of the brain has 1-14 per cent cholesterol while the grey matter has 60 per cent on dry basis. Sebum secreted by sebaceous gland also has large amounts of cholesterol.

Properties of Fat: The properties of fat are based on its composition and structure. Melting point of fat is the most important physical quality. Fats which have a high proportion of saturated fatty acids are solid at room temperature and unsaturated fatty acids are liquid at room temperature; If the melting point of a fat is above 50°C or 122° F it is poorly utilised by the body. The most important chemical properties of fats are hydrogenation, saponification, emulsification and rancidity. All these reactions are related to their chemical structure. The double bond linkages in the unsaturated fats make them readily available for all chemical reactions.

Iodine Value (number): Iodine value gives the number of grammes of iodine which combine with 100 grammes of fat Iodine value depends on the number of bonds it makes at the site of double bonds. This gives the degree of unsaturation of fats.

Saturated and Unsaturated Fatty Acids: Saturated fatty acids have all the carbon atoms in the chain saturated with hydrogen atoms. When a double bond is present between two carbon atoms the fatty acid is termed as unsaturated fatty acid. The degree of unsaturation varies according to the presence of double bonds. If one or more double bonds are present in a fat it is called unsaturated fatty acid. If there are more double bonds in a fat it is called polyunsaturated fatty acid.

Palmitic, stearic and butyric acids are saturated fatty acids. Oleic, linoleic, linolenic and arachidonic acids are unsaturated oils. They have one, two, three and four double bonds, respectively. The unsaturated fatty acids are often described as essential fatty acids. Vegetable oils are rich in linoleic acid and fish oils and animal fats are rich in arachidonic acid.

Saturated and Unsaturated Fatty Acids

Common name (saturated fatty acids)	*Molecular formula*	*Unsaturated fatty acids*	*Molecular formula*	*Double bonds (Nos)*
N-Butyric	C4 H8 02	Palmitoleic	C16 H30 02	1
Caproic	C6 H12 02	Oleic	C18 H34 02	1
Caprylic	C8 H16 02	Elaidic	C18 H34 02	1
Capric	CIOH20 02	Vaccenic	C18H34 02	1
Laurie	C12H24 02	Linoleic	C18H32 02	2
Myristk	C14H28 02	Linolenic	C18H30 02	3
Palmitic	C16H32 02	Ylinolenic	C18H30 02	3
Stearic	C18H36 02	Eleostearic	C18H30 02	3
Arachidic	C20H40 02	Arachidonic	C20H32 02	4

Percentage Composition of Some Fatty Acids in Common Oil and Fats

Fat	*Oleic acid (C18.1)*	*Linoleic acid (C18.2)*	*Polyunsaturated fatty acids*
Butter fat	3.3	4	2
Coconut oil	8	1.6	-
Cottonseed oil	30	56	2
Egg fact	52	9	3
Groundnut oil	65	17	Trace
Olive oil	65	15	Trace
Pig fat	47	4	3
Rapeseed oil	28	15	59
Red palmoil	52	5	Trace
Safflower seed oil	31	51	-
Sesame oil	45	41	Trace
Soyabean oil	29	51	6
Sunflower seed oil	14	73	1
Sardine oil	17.3	2.5	30.1

Some of the chemical reactions of fats are related to its unsaturated status.

Hydrogenation: Unsaturated fats have a tendency to combine with hydrogen at the site of its double bonds. Hydrogenation is such a process. Vegetable fats are exposed to hydrogen at high temperature in the presence of an alkali. Nickel or cobalt is used as a catalyst.

For example, groundnut, coconut, cottonseed and other oils are usually used for hydrogenation. On hydrogenation liquid fats turn into solids. Vanaspati is an example of hydrogenated fat from groundnut oil.

Rancidity: On keeping, fats develop an unpleasant odour and this is known as rancidity. There are two types of rancidity: hydrolytic and oxidative rancidity. One of the common reactions which cause the spoilage of fat is hydrolytic rancidity. In hydrolytic rancidity lipase hydrolyses fats into free fatty acids. Fatty acids with low molecular weight give an unpleasant odour to fats which is easily spoiled.

In oxidative variety unsaturated fatty acids are oxidised. The addition of oxygen may give rise to peroxides which on decomposition yield aldehydes and ketones. They have a foul odour. These peroxides are oxidising agents. They destroy important vitamins present in fats. This oxidation is prevented by the presence of antioxidants in the fat. Vegetables oils have tocopherols or vitamin E as a natural antioxidant in them. There are a b g (alpha, beta, gama) tocopherols. Among these alpha tocopherol is the most powerful antioxidant.

There are some other substances which can be used as antioxidants which give a prolonged shelf life to fats. Ascorbic acid or vitamin C, gum guaiac acid and propylgilate are some of them.

Saponification of Fats: The ester linkage of fatty acids with glycerol brings this process of saponification. Neutral fats are rich in ester linkage. When neutral fats are heated with sodium or potassium hydroxide the fat molecules may readily rupture at the ester linkage. It is thus hydrolysed into glycerol and sodium or potassium salts of fatty acids and this is known as soap. Thus, the formation of soap from the hydrolysis of fat with heat and alkali is termed as saponification. Saponification of fats, when performed under standard conditions, is used as an index to find out the molecular weight of fatty acids in a fat.

Emulsification: Fats and oils are insoluble in water because they are lighter than water.

When fat forms a homogeneous mixture with water it is called emulsification. This property is made use of in cookery.

Essential Fatty Acids: Nutritionally important fatty acids are known as essential fatty acids (EFA). Since the body cannot synthesise them the EFA must be supplied to the body through foods. Linoleic acids, linolenic acids and arachidonic acids are the essential fatty acids.

Functions of Fat in the Diet: The most important function of fat is due to its highest calorie density in foodstuffs. One gramme of fat supplies 9.3 kcal while 1 gramme of carbohydrate and protein supply 4.3 and 5.6 kcal respectively. Fats are the carriers of fat-soluble vitamins A, D, E and K. Fats also exert a thiamine-sparing action. Fats have a protein-sparing action also. Essential fatty acids help to maintain tissues, their normal structure and efficiency. Fats, especially phosphatides, form an essential part of nervous tissues. Adipose tissue, where fat is stored, serves as an insulation material in injury. Vital organs of the body are thus protected. In growth period it promotes growth. It also enables proper sexual maturity, especially in female experimental animals. Fat provides flavour and palatability to food.

Digestion and Metabolism of Fats: There is no digestive change for fat in the mouth. In the stomach a little fat is mechanically separated from other foodstuffs. Since carbohydrate and proteins are partially digested in the stomach fat is set free. In the stomach fat forms a coarse emulsion and is eventually released into the duodenum as small portions. Even though there is not much digestive action in the stomach for fat, digestion is delayed in the stomach by fat. The emulsified fat, known as chyme, takes three to four hours to transfer it from the stomach into the intestine.

Pancreatic lipase is the most powerful enzyme in fat digestion. This lipase breaks the alpha position of triglycerides and makes them free fatty acids and 1,2, diglycerides. 1,2 diglycerides are hydrolysed to 2, diglycerides and fatty acids. They combine with bile salts and emulsify fats. They reduce the surface tension of fat and make it available for enzyme activity. Bile salts promote emulsification and help the fat into partial solution of fats and fatty acids.

The water-soluble glycerol is quickly absorbed. Short chain fatty acids which are also soluble in water are transported into the portal blood. The products of fat digestion which reach the intestinal wall are resynthesised into new 'triglycerides by the intestine. In the jejunum triglycerides are absorbed. These triglycerides pass into the lymphatics in the form of small scattering particles called chylomicrons. Thorasic duct carries them into the systemic circulation.

Fats which are quickly absorbed are readily available for the tissues. If the fats remain in the gastrointestinal tract for a longer period for utilisation they add satiety value to the meal. Absorption of fat is based on its digestibility, melting point, the size of the fat molecules and its quantity. The age of a person is another factor which affects the absorption. Old people utilise fat slowly.

Lipids in Blood: In the post-absorptive state plasma contains about 500 mgs of total lipids/100 ml of blood. In this 180 mg/ 100 ml is cholesterol. Neutral fat occurs in blood plasma largely in the form of chylomicrons. In the liver triglycerides are resynthesised and they combine mainly with cholesterol and form beta lipoproteins and are passed into the circulation.

Cholesterol esters are converted to bile acids in the liver, re-excreted into the small intestine and reabsorbed. The concentration of bile may affect cholesterol synthesis in the intestine. Only forty percent of cholesterol in the circulating blood is from food.

The remaining portion is synthesised in the liver and intestine. Cholesterol synthesis in the body is controlled by various factors like calorie intake, cholesterol intake, fat intake and essential fatty acid content of the fat. Liver, skin, intestine, adrenal cortex, kidney and lungs are the organs which synthesise cholesterol. Acetate is the precursor of cholesterol and so all substances which yield acetate become precursor of cholesterol. Apart from fatty acids, carbohydrates, ethyl alcohol, alanine, and certain amino acids yield acetate or pyruvic acid or glycogen or glucose which can in turn form acetate. Thus every foodstuff forms acetate and cholesterol while calorie restriction and essential fatty acid deficiency inhibit hepatic cholesterol synthesis.

The adult human body synthesises about 2,000 mgs of cholesterol daily. The dietary intake of cholesterol ranges from 500 mg to 1,200 mg depending upon the consumption of milk, butter, eggs, meat and fish. Apart from these foods coconut oil and hydrogenated fat increase the blood cholesterol level. Unsaturated fats or essential fatty acids reduce the blood cholesterol level. Safflower oil, sesame oil, sunflower oil and soyabean oil are the unsaturated fats.

Metabolism: The metabolism of fatty acids and glycerol are separate. Glycerol enters the glycolytic pathway through the formation of 2 glycerophosphate. Fatty acids are oxidised to acetyl-CoA and it is again oxidized to Co_2 and water through Kreb's cycle of carbohydrate metabolism. Energy is the end product in the metabolism of fat also. In the stepwise reactions two carbon atoms are added or subtracted using co-enzyme A. For this reaction ATP or Adenosine triphosphate is required.

The enzyme which catalyzes the reaction is thiokinase. In the next step dehydrogenation of fatty acids takes place. Fatty acids are dehydrogenated into alpha beta unsaturated fatty acid. Alpha beta unsaturated fatty acids are converted or hydrated to hydroxyacyl-CoA. Hydroxy acid is oxidised to keto acid. The last step in the fatty acid oxidation is the thiolytic cleavage of Keteacyl-CoA ester to yield a fatty acid-CoA derivative. This is repeated and Acetyl-CoA molecules are released. Different fatty acids yield varying numbers of acetyl-CoA. For example, one molecule of steraic acid gives 9 molecules of acetyl-CoA and 1 molecule of palmitic acid gives 8 molecules of acetyl-CoA and 1 molecule of lauric acid gives 6 molecules of acetyl-CoA.

Acetyl-CoA thus formed is oxidised through Kreb's cycle and glycerol is converted into glyceraldehyde - 3 - phosphate and it enters into the glycolysis pathway.

Summary of Fat Metabolism

(1) Fatty acid - α β unsaturated fatty acid

+ H_2O

↓

(Beta) hydroxy fatty acid

Keto acid

↓

Acetyl CoA

(2) Glycerol ⟶ Glycerophosphate

↓

Glyceraldehyde - 3- phosphate.

Ketone Bodies: Normally acetyl-CoA combines with oxaloacetic acid which is oxidised in the TCA cycle. When large amounts of fat are metabolised in the absence of carbohydrate more acetyl-CoA is formed. When there is not enough oxaloacetic acid from carbohydrate, two molecules of acetyl-CoA condense to form acetoacetyl-CoA. This is hydrolysed into acetic acid and later converted into P-hydroxybutric acid and acetone. These metabolic end products of fats are known as ketone bodies.

When the concentration of ketone bodies in the blood increases the condition is known as ketosis. When it is excreted through urine it is called ketonuria. In normal condition acetoacetate is either oxidised directly to carbon dioxide and water or converted to two molecules of acetate which are later oxidised. In diabetes and during fasting ketone bodies are formed in excessive amounts. Apart from fats, amino acids like leuine also produce ketone bodies.

Lipo Proteins and Lipotropic Factors: Lipo proteins are lipids containing protein in blood and other parts of the body. The insoluble lipids when attached to proteins form soluble lipo proteins. Cholesterol, neutral fats and phospholipids bind them with protein to form lipo proteins.

Lipo protein in the liver along with choline enhances the oxidation of fatty acids. On diets which are high in fats with saturated fatty acids and which are low in protein or in choline there is an increased deposition of fat in liver. A high cholesterol diet also leads to the production of fatty liver. An increased choline intake will bring down the fat in the liver. This effect of choline on fat release is known as lipotropic action. Methionine likewise exerts a marked lipotropic action owing to its ability to promote the synthesis of choline by the transfer of methyl groups to suitable precursors. Inositol (B-complex vitamin), tryptophan and glutamic acid exert this lipotropic action. Essential fatty acid deficiency can also produce fatty liver, pantothenic acid or riboflavin deficiency may also lead to it

Requirements: The requirement of fats is about 50 to 60 gms for a normal person. The dietary fat is used for energy and it is also a source of essential fatty acids. At least fifty per cent of the fat must be from vegetable oils which are rich in essential fatty acids. An average Indian adult derives about 20 per cent of energy from dietary fat.

Sources: Vegetable oils, like groundnut oil, coconut oil, mustard oil or gingelly oil, hydrogenated oils, butter and ghee and lard are the common sources of fats.

Deficiency of Fats: During infancy a deficiency of essential fatty acids results in perianal irritation and skin changes like dry skin and itching. In adults phrynoderma is easily cured by essential fatty acids with vitamins B2 and A.

Different Material

Spices and Condiments: Spices and condiments are accessory foodstuffs mainly used for flavouring food. Some condiments are good to fair sources of carotene. Red chillies and coriander are examples of this. Green chillies are rich in vitamin C. Turmeric and tamarind are fair sources of iron. Garlic promotes intestinal synthesis of certain vitamins. Among the spices and condiments fenugreek seeds have the highest amount of protein, that is, 26.2 per cent. Cuminseeds, dry chillies and coriander seeds have a protein content of 18.7,15.9 and 14.1 per cent, respectively.

Caraway has a protein content of 17.1 and fat content of 21.8 per cent in it. Coriander, cuminseeds and dry pepper have a fat content of 16.1,15.0 and 6.8 per cent, respectively. Cuminseed is also rich in calcium. About 1.080 gm of calcium is present in 100 gms. Asafoetida and coriander are also good sources of calcium. Asafoetida, cuminseeds, cloves, fenugreek seeds, turmeric, tamarind and omum are good sources of iron. Cardamon, green chillies, coriander, cuminseeds, caraway and green pepper are rich sources of carotene. Dry chillies, nutmeg, cuminseeds, coriander and fenugreek seeds are good sources of thiamioe. Even though these condiments and spices are rich in vitamins and minerals they are used only in minute quantities. Black pepper, coriander, cumin, ferugreek, mustard seeds, chillies, nutmeg, cloves, cinnamon and cardamom are used as spices in our country.

The popularity of spices is due to their stimulating effect on appetite next to its flavour. Certain spices and condiments have medicinal effect. Turmeric is believed to be anti-inflammatory. In anti-allergic drugs turmeric is used. Turmeric also reduces blood cholesterol level. Consumption of spices and condiments in large quantities produces irritation in the mucous membranes of the alimentary canal. Peptic ulcer is common among people who use condiments and spices liberally. Herbs like curry leaves, coriander leaves, mint leaves and leaves of citrus fruits are used for preparing chutney powders and they are also used in many curries.

Gelatin Dishes: When the protein in connective tissues and bones—the collagen—is heated with water gelatin is obtained. To bring this change in collagen it has to be heated to 95°C. Some peptide bonds of collagen are hydrolysed to gel.

Commercially it is available in granulated, flake or sheet forms. It easily dissolves in hot water and on cooling it sets to a gel. Concentration of gelatin, cooling temperature, the presence of minerals like calcium and concentration of sugar affects the quality of the product. With a 2 to 3 per cent of gel concentration and at temperatures below 15°C a good gel is formed. Sugar in 6 to 10 per cent concentration enhances the speed of gel formation. If the gel is heated it melts and on cooling it sets again. Now vegetable gels are also available in the market.

Gelatin is used for decorating various types of desserts. For ice, souffles, baked ice-cream, or other creams, jelly trifles, caramel puddings, flans, milk puddings, sponge puddings, with stewed fruits, milk jellies, purees of fruits, junkets and with baked fruits gelatin can be used to make them more attractive. Gelatin acts as a stabiliser and reduces crystal formation of ice-creams.

Emulsifying and Stabilising Agents: Substances which, when added to food, gives a uniform dispersion of oils and fats in aqueous media are called emulsifiers or stabilising agents. In food industry it is often liberally used and they are harmful. Agar, edible-gums, brominated vegetable oils, gelatin, albumin, lecithin, starches, hydrolysed proteins and stearyl lastaric are some of the examples.

Sweetening Agents: Sugar, sugarcane juice, jaggery, molasses, honey, saccharin and cyclamates are the common agents used as sweetening.

Sugar (sucrose): Sugar is almost pure sucrose made up of glucose (dextrose) and fructose (laevulose). One gramme of sugar supplies 4 kcals and a teaspoonful (5 gms) provides 20 kcals. Brown sugar and white sugar are similar in nutritive value. Sweets made from sugar produce a sense of satiety and so they must be taken after a meal. If sweets are consumed at the beginning of a meal they reduce appetite. Sugar syrup is used as a preservative for fruits, jams and preserves. Excessive consumption of sugar and sweets affects the teeth adversely.

Sugarcane Juice: The juice obtained by crushing sugarcane is a refreshing drink. Addition of sugar and lemon juice improves its taste and flavour. Sugarcane is good for hepatitis patients. Jaggery is produced by heating sugarcane juice. It has been reputed to be good for the heart. Molasses is a byproduct in the manufacture of sugar. Fermentation of molasses produce alcohol.

Honey: Honey is a golden-coloured syrup made by bees from the nectar of flowers. It is a mixture of glucose and fructose. The quality of honey depends on the flower blossom. Honey is a soothing drink for patients with pharyngitis and tracheitis. Honey contains 17.7 per cent water; 36.5 per cent dextrose, 403 per cent laevulose and 1.9 per cent sucrose. Gur or Jaggery is obtained from sugarcane juice after processing. Palmyra palm, dale palm, or coconut palm is also used for it.

Saccharine and Cyclamates: Saccharine and cyclamates are used as sweetening agents for calorie-conscious people. Saccharine was discovered in 1879. It is 350 times sweeter than sugar and has a bitter taste in concentrated form. Cyclamate was discovered in 1937. It is 30 times sweeter than sucrose. Cyclamates are harmful to the body if consumed in large quantities. Malignant bladder tumours were produced in experimental animals. Excessive use of soft drinks with cyclamates produce renal tubular acidosis.

Leavening Agents: Leavening agents are substances which produce carbon dioxide in the product. This carbon dioxide contributes a porous appearance to the product as in bread, cake or in biscuits. Baker's yeast and baking powders are usually used as leavening agents. Baker's yeast when added to the dough, ferments sugar and produces carbon dioxide and alcohol. This gives the characteristic structure to the product. Baking powders are leavening agents produced by mixing an acid-reacting material and sodium bicarbonate with or without starch. There are three types of baking powders—fast acting, slow acting and double acting based on their speed of action.

Flavour Components in Foods: Food industry is a fast growing field. Much progress has been made in understanding the chemical nature of flavours in foods. Many artificial flavouring materials are now available in the market. The compounds which give the flavour characteristic in one food is given below:

Food		***Main flavour compound***
Banana	—	Isopentyl acetate
Lemon	—	Citral
Grape-Concord	—	Methyl anthranitrate
Potato	—	2-Isopropyo-3 methosy pyrazne
Beetroot	—	Geosmin
Apple	—	Ethyl -2- methyl butyrate

Use of Carbohydrates

Carbohydrates are compounds which contain carbon, hydrogen and oxygen. Oxygen and hydrogen are present in carbohydrates in the same proportion as in water. They are the main sources of energy for our body and the only source of energy for our nervous tissues.

Carbohydrates are mainly distributed among plant foods with few exceptions, like glycogen, lactose and ribose which are present in muscles or in liver, human milk and animal cells, respectively.

In various parts of the plants like leaves, stems, fruits, seeds and roots, carbohydrate is stored as starches and sugars. And the plant sources supply about 45 to 80 per cent of the energy requirement of the people throughout the world.

Plants through the process of photosynthesis manufacture carbohydrate and after its use store them as polysaccharides.

Importance: Carbohydrates are chemically known as saccharides, as carbon, hydrogen and oxygen form saccharide groups. Carbohydrates are classified, based on the number of saccharide groups in them:

a. Monosaccharides

b. Disaccharides, and

c. Polysaccharides.

Monosaccharides: On hydrolysis or by digestion, monosaccharides cannot be simplified as they are the simple form of carbohydrates. This is the absorbable form of carbohydrate in our body.

Based on the number of carbon atoms present in them, monosaccharides are grouped into triose (3 carbons), tetrose (4 carbons), pentose (5 carbons), and hexose (6 carbons).

Monosaccharides

Biose,	Triose	Tetrose
Pentose	Hexose	
$(C_2H_4O_2)$	$(C_3H_6O_3)$	$(C_4H_8O_4)$
$(C_5H_{10}O_5)$	$(C_6H_{12}O_6)$	
(*e.g.* Glycolical-	(*e.g.* Glyceral	(*e.g.* Erythrose)
(*e.g.* Xylose,	(*e.g.* Glucose,	
dehyde)	dehyde)	
ribose)	fructose)	

Biosetriose and tetrose are not nutritionally significant

Pentoses like ribose, xylose and arabinose are of little dietary significance, xylose and arabinose are widely distributed in many root vegetables and in many vegetables. Ribose is part of riboflavin and DNA and RNA Body can synthesise it and so it is not a dietary essential.

Xylose and Arabinose are not present in free state in nature. Both of them are present in gums of various origins like woodgum, cherry gum or gum arabic.

In human nutrition only hexoses are of importance. Commonly found hexoses are aldose and ketose. The three monosaccharides of hexose group are glucose or dextrose, fructose or levulose and galactose and mannose.

Glucose: Also known as grape sugar, glucose is an aldose sugar. It is a white crystalline substance easily soluble in water with a sweet taste. Glucose is readily absorbed from die stomach. Glucose is used to detect the disorders in carbohydrate metabolism because it is in the form of carbohydrate circulating in blood. A normal healthy person has 80-100 mgs of glucose/ 100 ml of blood. Glucose is present in fruits and honey.

Fructose known as fruit sugar is a ketosugar. It is sweeter than glucose. Fructose enters the cells without the aid of insulin and so it is recommended for diabetics. Honey is the richest source of fructose (30-40%). Fructose on reduction yields a mixture of sorbitol and mannitol.

Galactose is not present in nature. It is present only in milk sugar lactose; on hydrolysis lactose yields galactose and glucose. Galactose is present in cerebrosides in brain and nervous tissues.

Disaccharides: They are complex sugars with two saccharide groups and are formed by the condensation of two monosaccharides. On hydrolysis or toy digestion they are split into monosaccharides. Disaccharides have a structure of $C_{12}H_{22}O_{U}$. They are water soluble, diffusible and crystallisable. They vary in their sweetness. Commonly found disaccharides are sucrose, maltose and lactose.

Sucrose, otherwise known as invert sugar, is the commonest form of sugar in the diet. Either by hydrolysis or by enzyme action sucrose is simplified into glucose and fructose. It is present in sugarcane, beetroot and in many fruits and vegetables, and honey. Sucrose is formed by the condensation of one molecule of glucose and one molecule of fructose.

Maltose is a disaccharide which contains two molecules of glucose. It is otherwise known as malt sugar. Starch is converted to maltose before breaking down into glucose. It is present in sprouted grains. Cereal grains are rich in maltose.

Lactose is also known as milk sugar and contains glucose and galactose. Cow's milk contains about five per cent of lactose and human milk has about 6.8 per cent It is easily digestible but it is not as sweet as cane sugar. Lactose is synthesised in the mammary gland. It is less in sweetness compared to other sugars.

Trisaccharides: They are with three saccharide groups but are not of any nutritional significance. The fibrous parts of certain beans and root vegetables which are not digestible contain trisaccharides, *e.g.*, Raffinose is present in beetroots and cottonseed meal.

Tetrasaccharides: They contain four monosaccharides and like trisaccharides they are of no nutritional significance, *e.g.*, stachyose and scorodose.

Polysaccharides: They are large complex molecules of monosaccharides linked together. Several hundreds of glucose units are linked together in polysaccharides. They are insoluble in water. Starch, dextrins, glycogen, pectin and cellulose are the common forms of polysaccharides.

Starch: Plants store carbohydrates in the form of starch and it is the main source of nourishment for human race. Cereal grains, seeds, roots like potato, tapioca, yam and plantain contain considerable amount of starch. On cooking, starch absorbs water and it swells and ruptures.

This thickening quality of starch is used in cookery to produce a variety of dishes. Different sources of starch behave differently. Maize starch and cornflour are better 'thickening agents' than rice or wheat starch. All starches are broken down into glucose in the digestive system.

The characteristics of starch molecule depend upon the way in which 2,000 or more glucose units are linked. The number of glucose molecules in starch vary from 2,000 to 15,000. The glucose chain in starch are of two types. They are amylose and amylopectin.

(i) *Amylose:* Amylose is a long chain of glucose and starch has about 10-20 per cent of amylose molecule in it One molecule of amylose has 500-5,000 glucose molecules. Amylose is easily soluble in water and it is mainly responsible for the stiffening of cooked rice on keeping.

Amylose chain - G-G-G-G-G-G-G-G-G

(ii) *Amylopection:* Starch contains about 80-90 per cent of amylopectin in them. Amylopectin molecule has more glucose units. Usually one amylopectin molecule has

50,000-5,00,000 glucose molecule. Glucose molecules are arranged in branches.

```
              G
              G
              |
        G-G-G-G-G-G G
              G
              -G-G-G-G
              -G-G-G-G
      -G-G-G-G-G-G G
     G
     G
     G
     G
     G
```

Amylopectin molecule

Small branches of amylopectin has 20-30 glucose units and it has a colloidal property and this is responsible for thickening of starch water on heating.

Dextrins: When starch is partially broken into fragments either by digestion or by acids they are called dextrins. They are like starch itself. Dextrin is broken down into maltose. Starch is broken down into-amylodoxtrin, erythrodextrin, achrodextrin and maltose.

Pectin is a polysaccharide with no nutritional significance. But it is useful in the preparation of jam and jelly and thus it contributes to the palatability of foods.

Glycogen can be described as the animal starch as it is in this form animals store carbohydrate in the body. Glycogen is stored in the liver and muscles. This is the form of immediate energy for the body. About 350 gms carbohydrate is stored as glycogen in the body.

Cellulose is an insoluble, indigestible polysaccharide.

More than 3,000 glucose units are there in cellulose but it is not of human utilisation. Cattle can digest cellulose. Even though it is not of much food value it provides bulk to the diet and thus helps movements in the large intestine.

It prevents constipation and to an extent cancer of the bowel. It reduces the cholesterol level in blood as well as body weight. A high fibre diet can help in the treatment of obesity as it delays digestion and contributes satiety to obese people.

Sugar Alcohols: There are three sugar alcohols available commercially. They are sorbitol, mannitol and dulcitol. Sorbitol is used by diabetic patients.

D-sorbitol or D-glucitol is an alcohol made from glucose by hydrogenation. By this process the aldehyde (CHO) group of glucose is reduced to alcoholic groups (CH,OH). Mannitol is produced from mannose, a monosaccharide which does not occur free in nature.

Functions in Body

Energy Supply: The most important function of carbohydrate is to supply energy or fuel for the body. Energy from carbohydrate is utilised by the body for immediate use. Nervous tissues and erythrocytes use glucose for their energy whereas other tissues can make use of fatty acids for energy.

Special Functions of Carbohydrates in the Liver: Carbohydrates have protective and detoxifying action in liver, liver is protected against bacterial toxins by producing glycuronic acid or acetify groups from carbohydrates. They are excreted through kidneys. Since the defences of the liver against toxic agents are of great importance to the body, this function of the liver is very valuable.

For this purpose glycogen is made use of and thus carbohydrates play a significant role in removing poisonous substances from the liver. Carbohydrates, especially lactose, help the growth of desirable bacteria in the small intestine and help the synthesis of some B-complex vitamins in the intestinal tract. Lactose also helps in the absorption of calcium.

Carbohydrates and the Heart: Carbohydrate is used as the most efficient fuel for muscular exercise. Glycogen stored in the heart muscle is used for this purpose, especially in an emergency.

Carbohydrates and the Central Nervous System: Continuous supply of carbohydrate is essential for nervous tissues to function normally. Glucose as such has a specific influence in maintaining the functional integrity of nervous tissues.

Carbohydrates and Protein-sparing Effect: If carbohydrate supply is enough in the diet, protein is spared for important functions of it. This effect exerted by carbohydrate is called protein-sparing effect Insufficient intake of carbohydrate forces amino acids to be de-aminated and converted to fatty acids for energy purpose.

Carbohydrates and Non-essential Amino Acids: Carbon skeleton of carbohydrate is used by the body for the synthesis of non-essential amino acids.

Carbohydrates and Fats: Carbohydrate is essential for the oxidation of fat. In extreme restriction of carbohydrate, fat may be metabolised rapidly.

Carbohydrates in Diet: Carbohydrate provides flavour and variety to the diet Cellulose and other indigestible polysaccharides add bulk to the diet Carbohydrate also retains water content of colon which helps intestinal motility.

Rich Sources: The best sources of carbohydrates are cereals like rice, wheat, millets like ragi, maize, roots and tubers like potato, tapioca, sweet potato, yam and chalcosis, pulses, sugar and jaggery; honey, fruit and vegetables are the other sources of carbohydrates. Banana, apple, plantain and dried fruit are also good sources of carbohydrates.

In an average Indian diet carbohydrates are the cheap sources of calories. One gram of carbohydrate gives 4 kcals and an average Indian diet contains about 300 gms of carbohydrate.

Consumption and Digestion: In the mouth salivary digestion takes place. Carbohydrates are broken down by mastication and the saliva prepares the broken pieces of food for swallowing. Ptyalin, the enzyme present in saliva, acts on starch splitting it into dextrin and maltose.

This action continues in the stomach till the acidity of the gastric glands interferes. There is not any digestive action on carbohydrates in the stomach as there is no enzyme in the stomach to act on carbohydrates.

The main digestive action on carbohydrate takes place in the small intestine. Amylase of the pancreatic juice hydrolyses the remaining starch into maltose. The intestinal/ mucosa cells contain sucrose or invertose, maltose and lactose. These enzymes convert sucrose into glucose and fructose, maltose into glucose and glucose and lactose into glucose and galactode, respectively.

The end products of carbohydrate digestion—glucose, fructose and galactose—are absorbed in the intestine. Bulk of these sugars are absorbed in the duodenum and jejunum and small amounts are absorbed in the ileum. Glucose and galactose are readily absorbed while fructose is absorbed slowly. These monosaccharides are absorbed by the capillaries in the walls of the intestines and they are carried by the bloodstream to the liver and the tissues.

The absorption of carbohydrate from the intestine is controlled by certain factors like condition of the intestinal tract and muscle tone, endocrine glands like anterior pituitary, thyroid and adrenal cortex and their functions and vitamin B-complex content in the diet.

Carbohydrate Metabolism: Through portal circulation the absorbed monosaccharides are carried to the liver. There galactose and fructose are converted into glucose. The liver cells release some glucose into the bloodstream and blood carries it to the tissues. In the tissues this glucose is metabolised to release energy.

Excess of glucose is polymerised in the liver into glycogen and stored in the liver and muscles. This is again converted into glucose when energy is required. This process helps the bloodstream to maintain the normal blood sugar level. Glycogen is formed from glucose, fructose and galactose and it is shown below:

Glycogen Galactose

↑ ↓

←

Glucose-1 – Phosphate → Galactose-l-Phosphate

Glucose → Glucose-6-Phosphate

Fructose—-Fructose-6-Phosphate.

Formation of Glucose from other Monosaccharides

The carbohydrate metabolism in the body involves a series of chemical reactions. Glycogen formation, breakdown of glycogen and glucose to supply energy, formation of glucose from amino acid and glycerols of fat and formation of fat from carbohydrate are the main changes related to carbohydrate metabolism.

Breakdown of glycogen and glucose to yield energy involves a series of gradual reactions. Some of these chemical reactions take place in the absence of oxygen and it is known as the analrobic phase of carbohydrate metabolism. In the aerobic phase of metabolism oxygen is used. In total 38 ATP are yielded by each molecule of glucose: 8 ATP through glucolysis, 6 from pyruvic acid to Acetyl-CoA, 24 from TCA cycle.

Anaerobic Phase of Carbohydrate Metabolism (Glycolysis)

In glycolysis glycogen is degraded to glucose and it is converted to lactic acid. The energy release in the process of glycolysis is calculated from the breakdown of ATP (Adenosine Triphosphate) and regeneration of ATP. Two molecules of ATP are used and four molecules of ATP are formed. Glycolysis takes place in muscular tissues.

Glycolysis is also known as Embden-Meyerh of-Parnas scheme of carbohydrate metabolism. The basic steps involved in glycolysis are given in Fig.

Aerobic phase of carbohydrate metabolism involves oxidation of lactic acid into pyruvic acid and pyruvic acid to Acetyl-CoA. Acetyl-CoA undergoes a series of a chemical reactions whereby carbon dioxide, water and CoA are formed.

Glycogen

Glucose-1-Phosphate

Glucose Glucose-6—Phosphate

Fructose—6—Phosphate

Fructose-l-6-diphosphate

Glyceraldehyde-3-phosphate ——————>Dihydroxy acetone phosphate

1,3 Diphosphoglyceric acid

3 Phosphoglyceric acid

2 Phosphoglyceric acid

Pyruvic acid

Kreb cycle———————— Lactic acid

The conversion of Acetyl-CoA is otherwise known as Kreb's Tricarboxylic acid or T.CA. cycle.

Oxalacetic acid reacts with Acetyl-CoA and this is repeated.

Most of the absorbed glucose is oxidized releasing energy. During this process carbon dioxide and water are also formed. Some portion of glucose is converted to fatty acids and some amount combines with free amino group to form amino acids. B Vitamins like thiamine, niacin, riboflavin, pantothenic acid and B6 and minerals like magnesium and phosphorus are very essential for the proper metabolism of carbohydrates.

Certain hormones like epinephrine, thyroxine and glupxine and glucagon and islands of Langerhans are essential to maintain normal blood sugar level in a person. A healthy person's blood shows the fasting sugar level as 80-100 mgV 100 ml. of blood. After a meal it rises to 130-150 mgV 100 ml. of blood. If the blood sugar level is above this the person is termed as a diabetic patient

If the carbohydrate intake in the diet is inadequate, it leads to malnutrition and other metabolic disorders. Tissue proteins and fats will be utilised for energy purpose. If excess carbohydrate is taken, it will contribute towards obesity and diabetes.

Regulation of Blood Sugar: Blood circulates glucose continuously to each and every cell of the body as a source of energy and for the synthesis of a number of substances. From the blood if glucose is taken, the liver immediately replaces the blood glucose.

In the fasting state blood sugar level is 60-85 mgm per 100 ml of blood. Soon after a meal it rises to 140-150 mgm per 100 ml. If the body is in a good carbohydrate metabolic pathway this concentration will fall down to a normal level of 60-85 mgm.

The liver very efficiently maintains the normal blood sugar level. If the blood sugar concentration is high this condition is known as Hyperglycaemia. Hypoglycaemia occurs when the blood sugar level is below normal level. This occurs in certain abnormalities of liver function or when insulin is produced in excessive amounts by the pancreas.

The liver is the only organ in our body which can either

supply glucose to the circulation or to remove the excess sugar from blood.

Glucose is made available to blood from absorption, from intestine glucogenesis, *i.e.*, glycogen is produced from glucose and reconverts it; glucoseogenis, *i.e.*, de-aminated amino acids are used for the synthesis of glucose, glycogenolysis is reconversion of stored glycogen to glucose and lipogenesis, *i.e.*, excess glucose is converted to fatty acids and the use of glycerol to form glucose.

5

Food Fundamentals

This is not a special method of cookery but we have to use our knowledge of baking, boiling, steaming, etc., and in addition follow some rules which apply specially to food prepared for the sick. An invalid usually has a weak appetite and needs tempting because strength can only be regained through taking suitable food.

Wrong food may kill a patient, while badly prepared food will retard recovery. Good food, well cooked and served will do much towards saving the patient's life.

Convalescents vary in appetite according to the nature of their complaint. But whether they have good appetites or not, their food should be easily digestible as they cannot take as much exercise as a normal person.

Diet for the Diseased

(i) Consult the doctor about the patient's diet and obey his instructions.

(ii) Choose food that will supply necessary nourishment and be suitable to the illness,

(iii) Serve food which can be easily digested because in sickness the digestive system is often impaired.

(iv) Serve well-balanced meats, but pay special attention to the constituents according to the nature of the illness. In some cases meat, starches and sugar may have to be taken sparingly or even eliminated.

(v) In a serious illness, food is usually given in liquid form and should be as varied as possible,

(vi) In convalescence see that the solid food is light, easily digestible and served in small quantities,

(vii) As convalescence proceeds, increase the food in amount and variety.

(viii) Serve the best quality possible. Food should be absolutely fresh.

(ix) Vary the food and the method of cooking as much as possible. This stimulates appetite.

Proper Food

Liquid: Fresh fruit juice, barley water, beef tea, egg flip, lemonade or orangeade.

Soft Food: Made with milk, *e.g.*, junket or milk jelly, custard puddings (baked or steamed), blancmange, broths, light soups, rice conjee, etc.

Solid Food: White fish, fowl, rabbit, sweet bread or pancreas, liver dishes (on the advice of a doctor), khitchdi, etc.

Gathering Food

The fundamental principle in the storage of food is to keep it clean, cool and covered. Periodical inspection is a must, and there should be a regular turnover of goods. Daily cleaning, weekly cleaning and spring-cleaning are necessary to keep food free from dust and dirt.

Access to food stores must be restricted. No food should be kept in the kitchen or larder, apart from what is required during the day. The storage room should be dry, well-lit, ventilated, vermin-proof and clean.

Cupboards and shelves should be of simple design without dust-collecting ornamentation, unnecessary ledges or panels. Doors should be so designed as to enable the whole shelf to be clearly visible and accessible when doors are opened. Shelves must be constructed for specific purposes. They should be so fixed that they could be easily taken down and re-erected to facilitate regular and thorough cleaning. They should be narrow enough to enable all the goods to be easily accessible. Wide wooden shelves should not be made of planks which are joined, tongued or grooved but of 5 cm. x 1.25 cm. (2" x ½") slats laid with a clearance of 3.75 cm. to 5 cm. (1½"to 2") between each slat.

The one nearest to the wall should have a return board to form a lip. The back of the lip should be at least 5 cm. (2") from the wall, to leave room for the cleaning of the wall. Corners must be rounded, particularly where shelves on two walls meet at a corner, to avoid an uncleanable dust trap. As far as possible avoid having shelves above windows. Special shelving covered with marble, hard stone or tiles should be set aside for such materials as fat. One part of the shelf should be set aside for glass receptacles.

No foodstuff, even when in bins, boxes or cartons, should be stored directly on the floor. The bottom shelving should be at least 75 cm (26") above the floor. Containers for food should be provided, the nature of the foodstuff determining the nature of the receptacle required. These should be smooth, impervious, easily cleanable, resistant to wear denting, buckling, pitting, chipping, cracking and should withstand penetration by vermin, and the corrosive action of foods and cleaning compounds.

Stainless steel, model metal, vitrified china, glass or plastic should be used for moist unpackaged food or beverages, and galvanised steel for dry packaged and unpackaged foods. Solder must be nontoxic, and welded areas must be as resistant to corrosion as the parent metal. Food containers should stand clear of walls, should be kept scrupulously clean and unless the articles they contain are naturally wet, should be completely dry, both inside and outside.

Segregate food to be used promptly from the food that is to be held for some time. Use food in the order of priority—first in, first out.

Stocks should be inspected once a week. A stock list should be made to tick off inspection dates and to indicate conditions.

Keeping food clean not only means keeping it free from visible dirt, but also keeping it in such a way as to prevent the multiplication of germs High temperature kills them, and freezing preserves bacteria in suspended animation.

Hence food should be kept cool and dry. Perishables should be kept under refrigeration or, when no refrigeration is available, in the larder in closed containers or adequately protected with muslin or wire covers. Windows and doors should have fly-proof meshing.

Guide to Storage of Certain Fruits and Vegetables (R.H. 85 to 90%)

Commodity	*Temperature*		*Approximate Storage life (weeks)*	*Approximate post-storage life (days)"*
	°C	*°F*		
Apples	2°-3°	35°-38°	18	9
Apricots	-1°-0°	31°-32°	1-2	-
Bananas	11°-13°	52°-55°	3	1
Beans	0°-2°	32°-35°	2-3	-
Beetroot	0°-2°	32°-35°	6-8	-
Brinjal	8°-10°	47°-50°	3-4	-
Cashew apples	0°-2°	32°-35°	5	-
Cherries	-1°-0°	30°-32°	2-3	-
Carrots	0°-2°	32°-35°	13-17	-
Cabbage	0°-2°	32°-35°	9-13	-
Cauliflower	1°-2°	34°-35°	4-6	-

Contd...

Commodity	*Temperature* °C	°F	*Approximate Storage life (weeks)*	*Approximate post-storage life (days)"*
Guavas	8°-10°	47°-50°	4	-
Grapes	0°-2°	32°-35°	8	-
Grapefruit	5.5°-7°	42°-45°	8	3
Gooseberry (Indian)	0°-2°	32°-35°	8	3
Jackfruit	11°-13°	52°-55°	6	-
Lemons	5.5°-7°	42°-45°	6	3
Limes	8°-10°	47°-50°	8	-
Litchi	0°-2°	32°-35°	11	-
Mangoes:				
Alphonso	8°-10°	47°-50°	4	-
Badshapasand,				
Dasharj				
Dilpasand				
Khuddus				
Neelum				
RaspuriSafeda,				
Bangalore	5.5°-7°	42°-45°	4-6	3-4
Mangosteen	4°-5.5°	39°-42°	7	-
Oranges	5.5°-7°	42°-44°	18	-
Onions (R.H.70 to 75%)	0°-2°	32°-35°	17-26	-
Papaya	4°-5.5°	39°-42°	5	3
Passion fruit (purple)	5.5°-7°	42°-45°	3	-
Pears	-1°-0°	30°-32°	12	-
Persimmons	0°-2°	32°-35°	7	-
Peaches	-1°-0°	30°-32°	4-6	-
Peas	0°-2°	32°-35°	2-3	-
Pineapples	8°-10°	47°-50°	6	-
Plums	-1°-0°	30°-32°	2-4	-
Pomegranates	0°-2°	32°-35°	7-11	1
Pommelos	5.5°-7°	42°-45°	23	-

Contd...

Commodity	*Temperature* °C	°F	*Approximate Storage life (weeks)*	*Approximate post-storage life (days)"*
Potatoes	3°	37°-38°	26-35	-
Radishes	0°-2°	32°-35°	6-8	-
Sapota (Chikkoo)	0°-2°	32°-35°	10	5
Tomatoes				
(a) ripe	4.5°-7°	40°-45°	1-1½	-
(b) mature (turning to yellow)	13°-15.5°	55°-60°	3-4	-
Turnips	0°-2°	32°-35°	2-3	-

*at a temperature of 21°-34°C (70°-93°F)

Poultry: Dressed and ready to cook poultry deteriorates rapidly unless carefully handled and stored under properly controlled conditions. Freshly killed poultry, after dressing, should be drawn quickly, and wrapped loosely in waxed paper or aluminium foil and refrigerated. Such poultry can be kept for 2 to 3 days in a refrigerator at 3°C (about 38°F) or less.

Frozen poultry should be stored at -18°C (about 0°F) or less, and cooked soon after it has been defrosted. Generally speaking, frozen poultry should be used within 6 months after it has been processed to obtain satisfactory results.

Eggs: Eggs are highly perishable products and care must be taken in handling and storage. Eggs should always be stored in a refrigerator. They may lose as much of their fresh quality in 3 days at room temperature as in 2 weeks in the refrigerator.

Store with the large end up. At a range of temperatures and relative humidity varying from -2°C (about 28°F) and 85 per cent R.H. and 2°C (about 35° F) and 75 per cent R.H., eggs may be stored for as long as 9 to 10 months. Frozen eggs should be stored at -18° C (about 0° to 20°F).

Treating with oil or wax before refrigeration also extends shelf life for several months. A pure, colourless, odourless, tasteless mineral oil is used and the processing done under controlled conditions. Milk Bottled pasteurised milk can be safely kept for at least 3 to 4 days at refrigerated temperatures preferably between -1° and 4.4° C (about 34° and 40° F).

Know-how of up Keep: Knowing how to store food properly, and for how long you can keep it is very important: the food will be safe to eat; it will retain its flavour and texture; a high level of nutrients will be maintained; and you will not waste money on spoiled items.

Safety First: The three types of bacteria responsible for most cases of food poisoning (Salmonella, Staphylococcus and Streptococcus) are present around us all the time. Normally, they are not trouble some, but if given the right (actually wrong) conditions, bacteria can multiply in a matter of hours to dangerous levels.

The contributing factors to bacterial growth-temperatures between 1-52°C (about 50°-125°F) and time To thwart growth, do not be casual about handling food; always work with well-scrubbed utensils, cutting boards-and hands; cook or serve food as soon as possible after removing it from storage; refrigerate food immediately after the meal is over; always take precautionary care. It is often not possible to tell by taste or smell if food is contaminated.

Foods that Need Special Care: Give special attention to preparing and storming poultry, fish and sea food, meat, creamed mixtures, mayonnaise puddings and stuffings. Bacteria find these foods a good media to grow in. In particular: do not stuff poultry the night before; the cold stuffing may not heat up to a safe temperature when you cook the bird Never refrigerate cooked poultry with the stuffing in-store it separately If you have bought cracked eggs, use them only where they will be thoroughly cooked (baked or hard-boiled); Salmonella, which may be present on shells, could contaminate cracked eggs.

Know-how of Frozen Foods: It is safe to refreeze virtually all partially thawed foods if they still have ice crystals in them and are still firm in the centre. However, many foods (ice cream and uncooked baked goods for example) will not maintain top quality. Meat, fish and poultry purposely thawed in the refrigerator and kept no more than one day may be refrozen. However, do not refreeze combination dishes-pies stews, etc., that have been thawed.

With the exception of fruit and juice concentrates, foods thawed accidentally in the freezer over a period of days (because of a power failure, etc.) should not be refrozen, unless of course, they still have ice crystals. If food is completely thawed (on purpose or by accident), warmed to room temperature and left for more than two hours, throw it out.

The exceptions: fruit and juice concentrates, which ferment when spoiled. Throw away fruit if flavour is "off".

Pantry-shelf Storage Temperature: Store food in well ventilated cabinets—not over a cooking range or near refrigerator's exhaust. Use cool areas for storing large amounts of potatoes, onions, etc., and for long-term storage of canned foods.

Time: Though most staples and canned foods will keep indefinitely, buy no more than you expect to use in the recommended storage times given later. While foods will be safe beyond the recommended storage times, flavours will fade and textures will wilt. Date foods. Then check cabinets every six months and use up the oldest items.

Buying: Purchase the freshest-looking packages—messy or shop worn labels indicate old stock. Don't buy cans with swollen ends—the food has gone bad. Dented cans may be purchased, provided they are not punctured.

Home Canning: Use first-quality foods and the best techniques.

Temperature: From 1° to 4.5°C (about 34° to 40°F) is best. Above 4.5°C (40°F), food spoils rapidly. Check temperature with a refrigerator thermometer or an outdoor thermometer.

Time: Use foods quickly—do not depend on maximum storage time.

Wraps: Use foil, plastic wrap or bags, airtight containers. When meat, poultry or fish is bought in a plastic wrapped package, loosen ends of package to dry surface moisture—bacteria grow faster on moist surfaces.

General Care: Clean refrigerator regularly to cut down food odours. Remove spoiled foods immediately so that decay will not be passed on to other food.

Food	*Storage Time*	*Special Handling*
	STAPLES	
Baking powder	18 months	Keep covered and dry.
Bouillon cubes	1 year	Keep covered and dry.
Bread crumbs, dried	6 months	Keep covered and dry.
Cereals	4 months	Keep covered and dry.
Chocolate, premelted	1 year	Keep cool.
semisweet	2 years	Keep cool.
unsweetened	18 months	Keep cool.
Coffee, cans (unopened)	1 months	Refrigerate after opening.
Coffee, instant (opened)	2 months	Keep lid tightly closed.
(unopened)	6 months	
Condensed and evaporated	1 year	Refrigerate after opening.
milk Food	Storage Time	Special Handling
Flour (all types)	1 year	Put in airtight container.
Gelatine (all types)	18 months	Keep in original packets.
Honey, jams, syrups	1 year	Keep tightly covered.
Nonfat dry milk	6 months	Put in airtight container.
Pasta	2 years	Keep tightly closed.
Pudding mixes	1 year	Keep in original packets.
Rice, white	2 years +	Keep tightly closed.

Contd...

Food	Storage Time	Special Handling
Salad dressing (all types)	3 months	Refrigerate after opening.
Salad oil	1-3 months	
Shortening, solid	8 months	Refrigeration not needed.
Sugar, brown	4 months	Put in airtight container.
confectioners	4 months	Put in airtight container.
granulated, molasses	2 years +	Keep tightly covered.
Tea	2 years	Put in airtight container.
CANNED AND DRIED FOODS		
Fruits, canned	1 year	Keep in cool spot.
dried	6 months	Put in airtight container.
Gravies, canned	1 year	
Meat, fish, poultry, canned	1 year	
Pickles, olives	1 year	Refrigerate after opening.
Soups, canned, dried	1 year	Keep cool.
Vegetables, canned, dried	1 year	Keep cool.
HERBS, SPICES AND CONDIMENTS		
Ketchup (opened)	1 month	Herbs and spices
		Transfer from cartons to
whole spices	1 year	airtight containers, keep
ground spices	6 months	away from sunlight.
herbs	6 months	Check aroma; when it fades replace.
Tabasco, Worcestershire		2 years +
DAIRY PRODUCTS AT REFRIGERATED TEMPERATURES		
Butter, margarine	1-2 weeks	Keep tightly wrapped or covered. Hold only 2 day supply in butter keeper.
Buttermilk, sour cream	5 days to	Keep tightly covered. Turn
	or yogurt	
	2 weeks	unopened yogurt and sour cream containers upside down to prevent surface from drying. Once opened, store right-side up.
Cheese		Keep all cheese tightly pack
cottage	5 days to	aged in moisture-resistant
cream	2 weeks	wrap. If outside of hard
hard and wax-coated		cheese gets mouldy, just cut
cheeses-Cheddar, Edam,,		away mould—it would not

Contd...

Food	*Storage Time*	*Special Handling*
Gouda, Swiss, etc.		affect flavour. For longer
large pieces		storage, see freezer storage.
(unopened)	3-6 months	
(opened)	3—4 weeks	
(sliced)	2 weeks	Parmesan (grated)
	See pantry-chart. Does not need refrigeration.	
Processed (opened)	3-4 weeks	Unopened processed cheese
Cream-light, heavy,	3 days	need not be refrigerated
half-and-half		Keep tightly covered. Do not
Dips-sour cream, etc.		return unused cream to original containers. This would spread any bacteria present in leftover cream.
Commercial	2 weeks	Keep tightly covered.
homemade	2 days	Keep tightly covered.
Eggs	2 days	Keep small end of egg down,
in shells	3 weeks	to center yolks.
whites	3 days	Store in covered container.
yolks	3 days	Cover yolkes with water, cover container.
Milk evaporated (opened)		
pasteurised	4-5 days	Keep containers tightly closed Do not return unused milk
nonfat dry, skimmed	3-4 days	Remove entire lid of a can to make pouring easier Keep covered.
sweetened condensed	4-5 days.	remove entire lid of can to
	FRUITS AND VEGETABLES—FRESH	
Fruits		Sort fruit; discard bruised or
apples	1 week	decayed fruit. Do not wash
berries, cherries	1-2 days	before storing—moisture
citrus fruits	1 week	encourages spoilage. Store in
citrus juices		crisper or moisture resistant
bottled, reconstituted		bags or wrap. Keep fruit
frozen, canned	6 days	juices tightly covered.
melons	1 week	

Contd...

Food	*Storage Time*	*Special Handling*
other fruit	3-5 days	It is not necessary to remove canned fruit from can. Wrap uncut cantaloupe, honeydew, etc., to prevent odour spreading to other refrigerated food.
Vegetables		
beets, carrots, radishes	2 weeks	Remove leafy tops, keep in crisper.
mushrooms	1-2 days	Do not wash before storing.
onions, potatoes, sweet potatoes		Refrigeration not needed. See pantry-shelf chart.
shredded salad greens, cabbage	1-2 days	Keep in moisture-resistant wrap or bags.
unshelled peas, corn on husks, etc.	3-5 days	Keep in crisper or moisture-resistant wrap or bags.
other vegetables	3-5 days	Keep in crisper or moisture-resistant wrap or bags.
	MEAT, FISH AND POULTRY-FRESH UNCOOKED	
Meats		
beef, lamb, pork and veal chops	3—4 days	Store fresh meats loosely wrapped. Partial drying of surface increases keeping quality. If meat comes packaged in plastic wrap, loosen ends. Store in coldest part of refrigerator or in meat keeper.
ground meat	1-2 days	
roasts	5-6 days	
steaks	3-5 days	
stew meat	1-2 days	
variety meats (liver, heart, etc.)	1-2 days	
Fish and shellfish		
fresh cleaned fish including steaks and fillets	1 day	Store loosely wrapped. Keep in coldest part of refrigerator or in meat keeper.
Clams, crab, lobster in shell seafood including shucked clams, oysters, scallops, shrimps	2 days	Cook only live shellfish.
Poultry ready-to-cook chicken, duck or turkey	2 days	Store loosely wrapped. Keep fresh poultry in coldest part of refrigerator or in meat keeper.

Contd...

Food	*Storage Time*	*Special Handling*
	CURED AND SMOKED MEATS	
Bacon	5-7 days	Keep wrapped. Store in
Bologna	4—6 days	coldest part of refrigerator or
Corned beef	5-7 days	in meat keeper. Times given
Dried beef	10-12 days	are for opened packages of sliced
Dry and semi-dry sausages (salami, etc.)	2-3 weeks	sliced meats, Unopened vacuum packs keep about 2
Frankfurters	4-5 days	
Hams (Whole, halves)	1 week	
Hams, canned (unopened)	6 months	
Liver sausages	4-5 days	
Luncheon meat	3 days	
Sausages, fresh	2-3 days	
Sausages, smoked	2-3 days	
	CANNED FOOD AFTER OPENING	
Baby food	2-3 days	Store baby food covered.
Fish and seafood	1 day	Don't feed baby from jar;
Fruit	1 week	saliva may liquefy food.
Gravy, broths Meats	2 days	Store all canned foods tightly 2 days
covered. It is not necessary to		
Pickles, olives	1 month	remove food from can.
Poultry	1 day	
Sauce, tomato based	5 days	
Vegetables	3 days	
	OTHER FOODS	
Coffee	2 weeks	Keep tightly covered after opening
Honey, jams, jellies	2 weeks	Refrigeration not needed but storage life is lengthened if refrigerated.
Nuts	9 months	Refrigerate nuts after opening cans or packages. For longer storage, freeze.
Onions, potatoes sweet potatoes	2 weeks at room temperature	For longer storage, keep below 10°C (about 50°F), but not refrigerated. Keep dry, out of sun. Plan short storage in spring when sprouting is a serious problem.

Contd...

Food	*Storage Time*	*Special Handling.*
Parmesan cheese	2 months	Keep lid tightly closed.
Peanut butter		
unopened	9 months	Refrigeration not needed, but
after opening	2 months	it keeps longer if refrigerated.
Refrigerated biscuits, rolls pastries, refrigerated cookie dough	Expiry date on label	Products keep better if stored at the back of the refrigerator, where it is colder.
Soft drinks	3-6 months	
Syrups		Same as honey
Whipped topping mix	1 year	
Wines, table	2-3 days	Keep tightly closed.
cooking	2-3 months	Keep tightly closed.

Storing in Freezer

Temperature: -18°C (about 0°F) or below is best. The maximum temperature should be -15°C (about 5° F). Check temperature with freezer thermometer or outdoor thermometer, or use this rule of thumb: if the freezer does not keep ice cream brick solid, temperature is above the recommended level. If this is the case, do not plan to store food for more than a week or two.

Time: Date foods with an "expiry date" according to maximum storage time recommended below. Longer storage is not dangerous, but flavours and textures begin to deteriorate.

Wraps: Use foil, plastic bags and wraps, freezer wrap or freezer containers.

Commercially Frozen Foods: Do not buy them if they are battered, a sign that they may have been partially thawed. If foods have been partially thawed, then refrozen, use them within a few days.

Home-Frozen Foods: Use good-quality foods and proper techniques. Freeze foods quickly in coldest part of freezer, then store in another freezer area.

Food	***Storage Time***	***Special Handling***
	MEAT, FISH AND POULTRY	
Meat—home frozen and purchased frozen		
bacon	1 month	If meat is purchased fresh on
corned beef	2 weeks	trays and in plastic wrap,
frankfurters	1 month	check for holes. If none
minced beef, lamb and veal	4 months	freeze in this wrap for up to 1 month. For longer storage,
minced pork	3 months	wrap over with foil, plastic
ham slices	1 month	wrap or freezer wrap.
ham whole	2 months	Other meats should be

Contd...

Food	*Storage Time*	*Special Handling*
luncheon meats Roasts	1 month	wrapped as above. Make package as airtight as possible. Keep frankfurters, bacon, etc., Keep meat purchased frozen in original package, Thaw and cook according to sale instructions.
Steaks		
beef	1 year	
lamb	9 months	
pork	6 months	
veal	9 months	
Fish—home frozen and purchased frozen fillets and steaks from "lean" fish—cod, flounder,		To home-freeze fish, wrap in foil, plastic wrap or freezer
haddock, sole	6 months	wrap. Make package as air-tight as possible. Freeze in coldest part of freezer. For best results, prepare fish as directed in authoritative freezing guide. Keep fish purchased frozen in original wrapping. Thaw arid cook according to label directions.
"fatty" fish—bluefish,		
mackerel, perch, salmon	3 months	
breaded fish	3 months	
clams	3 months	
cooked fish or seafood	3 months	
king crab	10 months	
lobster tails	3 months	
oysters	4 months	
scallops	3 months	
shrimp, unbreaded	1 year	
shrimp, breaded	4 months	
Poultry—home frozen or purchased frozen		
chicken,		For home freezing, prepare as directed for fish and seafood (above). Thaw commercially frozen poultry according to label directions. Cook all thawed poultry within 1 day.
whole or cut-up parts	1 year	
chicken livers	3 months	
cooked poultry	3 months	
duck, turkey	6 months	

Contd...

Food	Storage Time	Special Handling
	FRUITS AND VEGETABLES	
Fruit-home frozen or purchased frozen		
berries, cherries,		
peaches, pears,		
pineapple, etc.	1 year	
(contd.)		
Food		
Storage Time		
Special Handling		
citrus fruit and juice		
—frozen at home	6 months	
fruit juice concentrate	1 year	
Vegetables—home frozen or purchased frozen		
home frozen	10 months	Cabbage, celery, salad greens, tomatoes do not freeze successfully.
purchased frozen-cartons,		
plastic bags or boil-in-bag	8 months	
	COMMERCIAL FROZEN FOODS	
Also see: meat, fish, poultry; fruits, vegetables; dairy products Baked goods		
breads, baked	3 months	Pick up frozen foods immediately before going to check-out counter. Place in home freezer as soon as possible. Purchase only foods frozen solid. Cook or thaw according to label instructions
breads, unbaked	2 months	
cakes		
cheesecake	3 months	
chocolate	4 months	
fruit cake	1 year	
sponge cake	2 months	
yellow or pound	6 months	
pies		
cream, custard	8 months	
fruit	8 months	
Main dishes		
meat, fish and poultry pies	3 months	
meat, fish and poultry casseroles	3 months	
TV dinners	6 months	
	HOME FROZEN FOODS	
Also see: meat, fish, poultry; fruits, vegetables; dairy products		
Breads	3 months	Package foods tightly in foil, plastic wrap, freezer or watertight freezer containers
Cakes	3 months	
Casserole dishes:		

Contd...

Food	*Storage Time*	*Special Handling*
meat, fish, poultry	3 months	For casseroles, allow
Cookies, dough, baked	3 months	room for expansion.
Nuts	3 months	Freese in coldest part of
Pies, unbaked	8 months	freezer.
	DAIRY PRODUCTS	
Butter, margarine	9 months	Store in airtight freezer container or wrapped in freezer wrap.
Buttermilk		
sour cream or yogurt		Do not freeze.
Cheese		
Camembert	3 months	Thaw in refrigerator.
Cottage, farmers' cheese (dry curd only)	3 months	Thaw in refrigerator. Do not freeze creamed cottage cheese—it gets mushy.
Cream cheese,		Thaw in refrigerator.
Neufchatel	1 month	Cut and wrap cheese in small
Hard cheeses—Cheddar, Edam, Gouda, Swiss, Brick, etc.	3 months	pieces for freezer storage. When frozen, may show mottled colour due to surface moisture. Thaw in refrigerator.
Processed	3 months	
Roquefort, Blue	3 months	Becomes crumbly after thawing Still fine for salads and melting.
Cream—light, heavy. half-and-half	2 months	Heavy cream may not whip after thawing. Use cream for cooking. Thaw in refrigerator.
Cream, whipped	1 month	Make dollops of whipped cream; freeze firm. Place in plastic bag or carton; seal; store in freezer.
Place on top of dessert to thaw.		
Eggs		
in shells		
whites	1 year	Store in covered container. Freeze in amounts for favourite recipes.
yolks	1 year	For sweet dishes: Mix one cup yolks with 1 tablespoon corn syrup or sugar. For other cooking: substitute ½ teaspoon salt for sugar.

Contd...

Food	*Storage Time*	*Special Handling*
Ice cream, ice milk, sherbet	1 month	Cover surface with plastic wrap or foil after each use to keep from drying out.
Milk	3 months	Freezing affects flavour and appearance. Use in cooking and baking. Allow room for expansion in freezer container. Thaw in refrigerator.

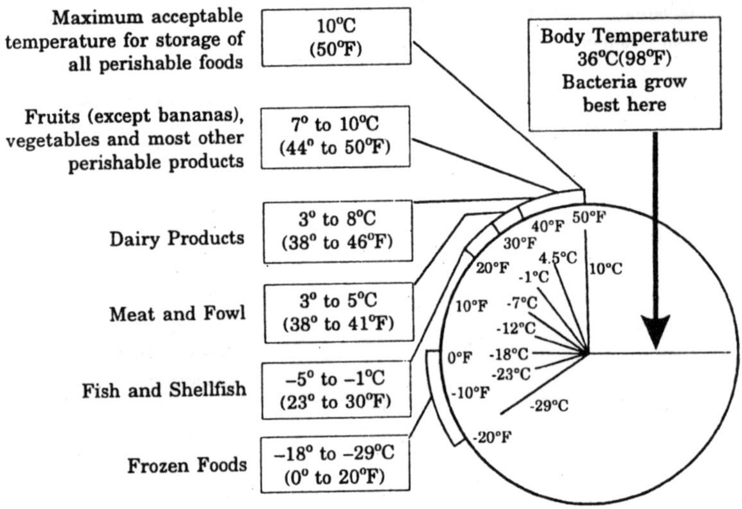

6

Food Ingredients

Use of Fish

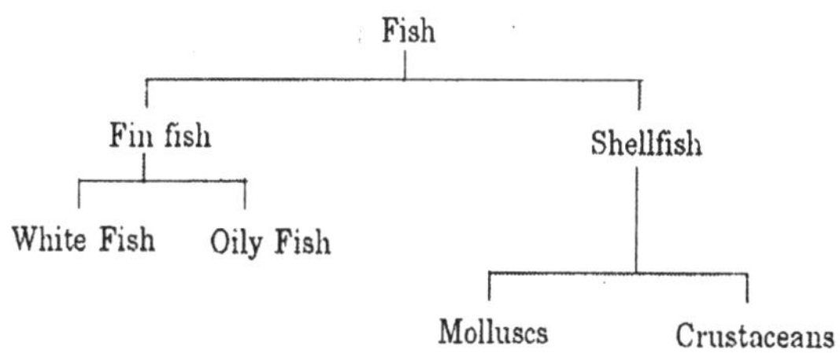

Fin fish are vertebrates and have skin and scales which cover the body. They move with the help of fins. They are subdivided into: (1) *White fish* which are mainly flat fish, and contain oil only in the liver. Most of these are deep sea fish. Common local examples are pomfret, sole, etc. (2) *Oily fish* which are mainly round fish and contain fat all over the body. The amount of fat varies from 1.5% to 20% in different varieties. These fish are often pigmented and tend to be surface fish. Example: mackerels, sardines, etc.

Shellfish, as the name denotes, have a shell covering the body. They are invertebrates. They are subdivided into:

(1) *Molluscs*: (a) Bivalves (*e.g.* oysters, scallops, mussels, clams and cockles) which have two distinctly separate shells

joined by a hinge-like membrane. The movements of the shell are controlled by a strong muscle. When the muscle relaxes, the two halves of the shell fall open. The shell also opens when the organism dies, thus exposing the contents of the shell to contamination from outside, resulting in quick putrefaction. (b) *Univalves:* (*e.g.* whelks and winkles). These are recognised by the characteristic spiral formation of their shells, which unlike those of bivalves are not divided into halves.The shells of molluscs increase at the rate of one ring per year to allow for the growth of the organism. The age of the mollusc can thus be roughly estimated by the number of rings on the shell.

(2) Crustaceans have a segmented, crust-like shell (*e.g.* lobsters, crabs, prawns, shrimps). The shells of crustaceans do not grow with the fish, unlike those of the molluscs, but are shed every year, with a new one forming to suit their new size.

How to Select Fish: The following points must be borne in mind when selecting fish: (i) Eyes should be bright and not sunken, (ii) Gills should be red, (iii) The tail should be stiff, (iv) The flesh should be firm and not flabby, (v) The scales if any should be plentiful, (vi) There should be no unpleasant odour, (vii) To test a cut piece, press down with a finger, and if an impression is left then the fish is stale, (viii) Any tendency for the raw flesh to come away from the bones is a dangerous sign.

Scaling and Cleaning:

(i) To scale, use the blunt side of the knife. Holding the fish by the tail, scrape firmly from the tail towards the head. Wash well to remove any loose scales,

(ii) Remove fins and trim tail,

(iii) To clean fish such as herrings, mackerel, etc., slit the belly, using a pair of kitchen scissors, from the head to almost half way down. Remove the head if desired,

(iv) In the case of flat fish such as pomfret, etc., cut away the gills, and make a small opening in the belly. Remove the entrails and wash well. If the head is to be removed, do this with a semicircular cut at the base of the head.

To fillet the fish: Slit the fish down the backbone with a sharp knife and separate the flesh from the backbone on each side with short sharp strokes. Then slit from the side and take out each fillet.

To bone herrings After slitting and cleaning, open gently. Turn over and place the inside on hard board. Press firmly along the backbone to loosen the bone. Turn over and starting at the head raise the backbone with the thumb and forefinger. Pull the bone steadily away from the fish. After pulling out the bone, rinse the fish well and drain.

Cleaning of Fish

What to do	*What to do*
Do not work without having a basin prepared to hold the fish bones, and another one for the offal.	Wash and dry fish with a clean swab. Cut off all the fins and trim the tail; re move the eyes if necessary.
Do not work directly on a table; use a chopping board.	Remove the scales, if any, by scraping them with a knife from tail to head, then remove the gills.
Do not throw the fish scales on to the table; remove them from the knife by warping the knife against the side of the basin to collect them. Don't mix different fish together; always begin with one kind and place them on one side to be washed with the others, when they have all been cleaned from inside.	Remove all the entrails, roe, liver etc., from the gill-slits by pulling the gut out with a fork, or the hook of a small ladle handle. Make a small opening, to 5 cm. long, from the vent to the belly of fish such as herring, whiting, trout, and salmon, to remove the roe and clean the fish properly. For turbot, brill, etc. make the opening behind the head.
Do not have your table covered with scales or offal of fish. Keep your chopping-board and table clean by swabbing often.	After removing roe, such as herring roe, place them immediately in a basin of cold water.
Do not forget, when removing the entails of a fish, to clear the bones of the blood vessels; and ensure that there is nothing left inside the fish by feeling with one finger.	If the head has to be removed, cut it from the gill in a V-shape.
Do not cut the ligament between the head and the body when removing the gills from small fish. Do not burst open or cut off the belly of a small fish when cleaning the inside	Wash the inside of the fish well under running cold water. With a clean or clothe swab or cloth dry inside as well. Then inspect the interior of the fish carefully.
Do not place a fish on a tray unless it has been well washed, dried and carefully examined.	

Cooking of Fish: The flesh of fish is very delicate and, therefore, great care must be taken to prevent the flesh from breaking during cooking. When boiling or cooking fish, the liquid should be at simmering point before fish is added as the connective tissue in fish is softer and the quantity is less than in meat. When preparing fish stew or fish curry, fish is put into cold liquid to flavour the gravy as well.

Whole fish may also be covered with a cold liquid and brought to boil. This is done to prevent shrinkage of the skin. Steaming is a better method of cooking fish as it is a gentle method. Fish cookery is not a special method of cooking. All methods of cooking can be applied bearing in mind that the delicacy of the flesh of the fish and the fat content of the fish determine the cooking method.

The time taken is always much less than for cooking meat. An average timetable is as follows, but much depends on the size and shape of the fish.

Boiling and baking		10 minutes per 500 gm and 10 minutes over
Steaming		15 minutes per 500 gm and 15 minutes over.
Stewing		Half to three quarter of an hour formes time the stew begins to simmer.
Frying (for small fish or thin pieces)		5 minutes
	Grilling	5 to 7 minute's
Lean fish	Steamed or	Poached, deep-fried or baked in a
Oily fish	Baked	Sauce. Baked, grilled or pan-fried.

Boiling of fish: Fish is generally boiled in a court bouillon. This is made of salt, water, milk and lemon juice for sea fish with white flesh. For salmon, trout and shellfish and various types of river fish the following court bouillon is used.

Court Bouillon

Water	1 litre
Salt	15 gm
Vinegar	75 ml
Carrots (sliced)	60 gm

Onion (sliced)	60 gm
Bay leaf	1
Parsley stalks	2 to 3
Peppercorns	6
Thyme	1

Simmer all ingredients for 30 to 40 minutes.

Strain and use as required.

The stock after fish is boiled can be used as a base for different types of sauces. Fish is cooked when the flesh separates easily from the bone. Test at the thickest part nearer the bone. It there is no bone then the fish is considered done when a creamy substance begins to run from the fish.

Steaming of Fish: Place the fish in a steamer. Sprinkle with salt and pepper. The water in the steamer must be fast boiling all the time the fish is cooking, if large, the fish can be turned once. Test as for boiling.

Grilling of Fish: Heat the grill and grease the rack. Clean fish. Sprinkle over with salt, pepper and lime juice. Gashes should be made on the fish to allow heat to penetrate, otherwise the outside becomes dry before the inside is cooked. For large fish, baste frequently with melted fat. For oily fish such as herrings and sardines, no extra fat is required.

Frying of Fish: Fish can be fried whole, if small, or cut into steaks or fillets or pieces if large. For cut pieces, it is preferable to coat the fish with egg and bread crumbs or seasoned flour and milk.

Baking of Fish: This method is most suitable for medium sized fish or the middle portion of a large fish. Scale and clean, but leave the head and tail on. Stuff the belly with a savoury forcemeat (11 gm fresh bread crumbs, 60 gm fat, grated rind of half a lime, 2 tsp chopped parsley, ½ tsp mixed herbs, salt and pepper and egg or milk to bind). Sew up with a needle and coarse thread. Do not fill the fish too tightly or it will burst, since the stuffing swells when it is cooked. Place the fish in a baking tray with a little fat, in a moderately hot oven, and cook till the flesh leaves the bones. To serve, remove the thread. Garnish with parsley and lemon. Serve with a sharp sauce.

Soused Herring or Mackerel: Soused herrings and mackerel are served as an hors d'oeuvre or for breakfast. Scale and clean the fish. Cut off the head and tail. Remove entrails. Wash well. Put back roe if any. Place on a plate. Sprinkle over with a little salt and leave for a couple of hours. Drain the fish. Roll up and place in a casserole dish. To 500 gm of fish add 12 peppercorns, 1 bay leaf and a large onion thinly sliced. Pour over equal proportions of vinegar and water, the liquid being just sufficient to cover the fish. Bring quickly to boil and simmer for a few minutes before removing from the casserole. Serve garnished with parsley.

Shellfish

Choosing Shellfish: They should be bought in season as far as possible. They should be medium- sized and of good weight. It is best to buy them' alive, particularly oysters, crabs, and lobsters. If they are dead, the following precautions must be taken: (a) The claws of crabs should be springy and not hanging down, (b) The eyes should be bright, (c) The tail of lobsters should spring back when stretched out. (d) Shrimps and prawns must be crisp, (e) Oyster shells should be tightly closed.

Preparing and Serving Shellfish: Shellfish can be served in a variety of ways. Lobsters and crabs are boiled alive. The meat is removed after cooking, flaked and replaced in their own shells with sauces or served in salads or other cold dishes. The meat is mixed with sauce and served in patties, or in scallop shells or on toast and canapes. Smaller shellfish such as prawns and shrimps must be shelled first, the intestines (the black thread- like substance found at the back) removed, then washed and cooked. They are served whole, boiled or steamed, stewed or in curries, fried with masala or in batter, devilled and in soups. Shrimps and oysters can be covered with sauces. Lemon and vinegar are served with shellfish to aid in digestion.

What to Serve with Fish: When fish is served as a main dish, it is usually served with potatoes, a salad or a cooked vegetable and a sauce. Here are some suggestions.

Fish	*Potatoes*	*Vegetables*	*Accompaniments*
Fried or grilled	Fried, boiled or sauteed	Green salad, cole slaw, green peas or beans, carrots, tomatoes parsnips, cauliflower,	Hollandaise, tartare, caper, tomato, mustard or cucumber sauces, lemon, parsley, butter, chutney, pickled beetroot or red cabbage.
Boiled or steamed	Boiled in jacket or mashed	Green peas or beans, carrots, tomatoes, spinach, celery, green salad, cucumber salad, parsnips, hot beetroot, sprouts, cauliflower, onions.	Hollandaise, cheese, horseradish, chutney, parsley, egg, fennel, anchovy or lemon sauces.
Baked	Baked, boiled or sauteed	Green peas or beans, carrots, green salad, tomatoes, cucumber salad, spinach, celery, parsnips, hot beetroot, sprouts, cauliflower, Onions.	Hollandaise, mustard, anchovy, cucumber, tomato, tartare, caper or brown sauces, lemon, parsley, butter.

Meat: The term meat is generally applied only to butcher's meat—beef, mutton, lamb, pork and kid.

The tenderness of the meat depends on the age and feeding of the animal, on the hanging and preparation after killing. Muscle structures have an effect on the tenderness and the texture of the meat. The skeletal muscles are made up of fibres which are composed of proteins, salts and nitrogenous extractives. The fibres are grouped parallel to each other in bundles held together by connective tissue and surrounded by a sheath of heavier connective tissue. The size of the bundles varies with different muscles and determines to some extent the texture of the meat. There are two kinds of connective tissue, collagenous or white connective tissue and elastic or yellow connective tissue.

The flavour of meat is determined by various factors such as feeding, age, sex and maturity. The older the flesh the greater the flavour in most cases; young flesh is not so highly flavoured but is sweeter and more tender. Flesh of male animals is stronger in flavour than that of females unless the animal has been castrated. Well developed muscles are usually more strongly flavoured, *e.g.*, chicken leg as compared with breast. The longer the meat is hung the stronger the flavour, owing to certain chemical changes. Meat should not be eaten immediately after killing.

It has to be hung till tender. Shortly after death rigor mortis sets in and the muscles stiffen. Meat must be hung at 1° to 3°C (about 34° to 38°F) at least, till this passes off and until the acids which then develop have had a chance to soften the connective tissue. The enzyme present in the tissues also helps to make the meat render and more juicy.

In cold countries this process takes at least 3 days but it is much quicker in warmer temperatures, varying between 12 to 24 hours. If so desired meat can be hung for a longer period. Not all meats should be hung like this and veal and pork particularly should be eaten fresh. They should also be thoroughly cooked before use.

Generally speaking, old meat and meat with considerable muscular development will be darker in colour and stronger in flavour. The flavour of the meat varies depending on the saltiness and sweetness of the blood, the species, the age and sex of the animal and the degree of maturity when eaten. Another factor is fat. Pork and mutton for example, have individual flavours depending on the fat. The third factor is the amount of sugar in the meat which on cooking tends to caramelise. Wherever possible select meat which is marbled. This is meat where the fat is interspersed between the muscle fibres and connective tissue. This shows good feeding. Marbling helps to keep the meat moist when cooked by dry methods. Marbled meat is also more nutritious.

There are no special methods of cooking meat and the fundamental methods of cooking can be deftly adapted to suit the particular joint. The amount of connective tissue present in the meat determines to a great extent the method of cooking. When connective tissue is present in large amounts, the meat, being tough, is cooked by moist methods.

For those cuts which have only a small amount of connective tissue, dry heat cooking may be used. Long-fired cuts of meat require longer cooking than short-fired ones. They should be stewed or braised to make them tender. Shorter-fired meat is cooked by dry heat. The most developed muscle fibres are usually the longest and these coarser fibres are found in the legs, neck and tail. The shorter fibres are found in the fillet, rump and loin.

Dry heat cooking (roasting, frying and grilling or broiling) develops the characteristic meat flavour, coagulates the proteins, thus making fibres firm, hardens the connective tissue, and changes the colour. This change in colour is caused by a change in the composition of haemoglobin. Shrinkage is observed in meat cooked by the dry heat method. Moist cooking (braising, boiling, stewing, etc.) besides developing flavour and coagulating the meat, softens the collagenous tissue, thus rendering meat tender. Elastic connective tissue is softened only slightly by moist heat. There is no shrinking in meat cooked by moist methods.

When cooking meat it is desirable to retain the juices in the fibres. Steaks, roasts, etc., are cooked either at a constant temperature of 149°C (about 300°F) for beef, veal and lamb and 177°C (about 350°F) for pork, or by the searing method, where the meat is cooked at a high temperature for the first 10 minutes and then cooked at a lowered temperature. This method was supposed to help in retention of juices but this has been disproved although the aroma, flavour and colour that accompany searing are preferred by some. Steaks are usually grilled at 204.5°C (about 400°F). Studies have illustrated that a lower oven temperature 171°-177°C (about 300°-350°F) as compared to a higher temperature 218.3°-232.2°C (about 425°-450°F) results in less drip loss, less shrinkage, increased juiciness and more uniform colour throughout a cut. Overcooking or extended cooking at high temperature denatures the protein of meat, eggs, fish and birds or they toughen or become stringy or all three.

If a roast is cooked at a high temperature, the time of cooking is decreased, but dripping and evaporation losses are higher, and the meat is not so evenly cooked. The degree of 'doneness' also affects this loss, the steak done 'rare' or 'underdone' showing' less than that done 'medium' or Veil done'. A meat thermometer helps in determining the degree of 'doneness'. The thermometer should be inserted so that the bulb is at the centre of the cut but not in contact with the bone. For thin cuts the thermometer is not very satisfactory.

Meat meant for boiling should be dipped into salted boiling water and then allowed to simmer till tender. A pressure saucepan may be used to save time when cooking less tender cuts of meat. Care should be taken to prevent cooking of meat in a pressure cooker or at a high temperature longer than necessary, otherwise it becomes hard, dry and flavorless.

Frozen meat may be prepared by the same methods as fresh meat. It need not be thawed before cooking. If it is not thawed, however, the cooking time is considerably increased, to as much as two to three times the time required for cooking thawed or fresh meat. Thawing can be done in the refrigerator or at room temperature. When meat is not thawed, it does not retain a coating such as egg and bread crumbs.

Tenderisers such as papain (taken from raw papaya) may be used to soften the fibres before cooking. These must be applied to the meat, which should then be allowed to stand for one hour for a cut of 2.5 cm. (l") thickness. The acidity of vinegar, lemon juice, curd, cider and wine which can be added before or during cooking also helps to tenderise meat.

Sign of Quality: Mutton and lamb lamb is meat from a lamb under one year old):

(i) A good quality animal should be compact and evenly fleshed,

(ii) The lean flesh should be firm, of a pleasing dull red colour and of a fine texture or grain,

(iii) There should be an even distribution of surface fat, which should be hard, brittle and flaky in structure, and of a clear white colour,

(iv) In a young animal the bone is pink and porous, so that when it is cut, a small amount of blood can be seen. With progressive aging the bones become hard, dense, white and inclined to splinter when chopped.

Order of Dissection:

(i) Remove the shoulders,

(ii) Remove the breasts,

(iii) Remove the middle neck and scrag end.

(iv) Remove the legs.

(v) Divide the saddle from the best end.

Approximate weight of the carcass of an Indian animal:

Lamb: 10 kg. (22 1b.)

Mutton: 16.8 kg. (37 lb.)

Whereas the approximate weight of an English animal is:

Lamb: 14 kg. (32 lb.)

Mutton: 23 kg. (50 lb.)

Pork

Fresh pork is at its best from September to April (the months with an 'r' in them). At 5 to 6 weeks it is known as suckling pig.

Sign of quality

(i) Lean flesh should be pale pink, firm and of a fine texture,

(ii) The fat should be white, firm, smooth and not excessive,

(iii) Bones should be small, fine and pinkish,

(iv) The skin or rind should be smooth.

Joints and their Uses

	Joints	*Uses*		*Approx. Weight*	
			French	(Lamb)	Mutton)
(i)	Shoulder	(L'epaule)	Roasting Stewing	2.00 kg	3.5 kg.
(ii)	Leg	(Le gigot)	Roasting Boiling	2.75 kg.	3.75 kg.
(iii)	Breast	(La poitrine)	Roasting Stewing	0.9 kg	1.75 kg.
(iv)	Middle neck	(Le collet)	Stewing	1.15 kg.	1.7 kg.
(v)	Scrag end	(Le cote decourverte)	Stewing Broths	450 gm.	0.9 kg.
(vi)	Best end	(Le carre)	Roasting Grilling Frying	575 gm.	20 kg.

Contd...

	Joints	*Uses*		*Approx. Weight*	
			French	(Lamb)	Mutton)
(vii)	Saddle	(La selle)	Roasting Grilling Frying	1 kg.	3.5 kg.
(viii)	Kidney	(Le rognon)	Grilling Saute Braising	1.5 kg.	2.25 kg.
(ix)	Heart	(Le coeur)	Braising	115 gm	60 gm.
(x)	Liver	(Le foie)	Frying		350 gm.
(xi)	Sweetbread	(Le ris de veau)	Braising Frying		
(xii)	Tongue	(La langue)	Boiling Braising		

Pork: Fresh pork is at its best from September to April (the months with an 'r' in them). At 5 to 6 weeks it is known as suckling pig.

Sign of Quality:

(i) Lean flesh should be pale pink, firm and of a fine texture.

(ii) The fat should be white, firm, smooth and not excessive.

(iii) Bones should be small, fine and pinkish.

(iv) The skin or rind should be smooth.

1. Leg
2. Loin
3. Spare-rib
4. Belly
5. Shoulder
6. Head (whole)

   Trotters

   Kidneys

   Liver

Order of dissection:

(i) Remove the head.

(ii) Remove the trotters,

(iii) Remove the leg.

(iv) Remove the shoulder.

(v) Remove spare ribs.

(vi) Divide loin from the belly.

Joints and their Uses

	JOINTS French	*USES*	*APPROX. WEIGHT*
(i) Thigh	(La cuisse) Boiling	Roasting	4.55 kg.
(ii) Loin	(La longe) Frying Grilling	Roasting	5.45 kg.
(iii) Spare rib verte de pore)	(Le cote decou- Pies	Roasting	1.35 kg.
(iv) Belly (Breast)	(La poitrine) Boiling	Pickling	1.8 kg.

Contd...

		JOINTS *French*	*USES*	*APPROX. WEIGHT*
(v)	Shoulder	(L'epaule) Sausages Pies	Roasting	2.7 kg.
(vi)	Head (whole)	(La tete)	Brawn	3.6
(vii)	Trotters	(Le pied) Boiling	Grilling	
(viii)	Kidney	(Le rognon) Grilling	Saute	
(ix)	Liver	(Le foie)	Pate	

Bacon and Ham: Bacon is made by preserving the meat of carefully bred pigs. The preservative used is salt. This can be applied in two ways:

(i) by covering the meat with dry salt and allowing at least 4 weeks penetration;

(ii) by using a salt solution (brine).

This is injected into the flesh and then the meat is immersed in a salt solution for 5 days. Sometimes a combination of the two methods is used—brine is injected into the meat and the sides are then stacked and covered with dry salt for 2 to 3 weeks. It is then matured for about 3 weeks under controlled temperature. Desirable bacteria acting on the meat juices and salt produce the characteristics associated with well matured bacon.

This bacon is now ready for sale as 'green' or 'pale' bacon. In certain parts of the world the bacon is preferred smoked. So it is then hung over smouldering wood dust for 2 days. This extracts excessive moisture and gives the bacon a distinctive flavour. Smoking also improves the preservation as it acts as an antiseptic. This is particularly suitable in hot climates.

Good quality, well cured bacon should have a pleasant

smell; the rind should be thin and smooth, the fat firm and free from any yellow marks, the lean part a good deep pink colour and the flavour mild and mellow.

1. End Collar
2. Prime Collar
3. Fore Hock
4. Top Back
5. Top Streaky
6. Back and Ribs
7. Prime Streaky
8. Short Back
9. Thin Streaky
10. Long Back
11. Flank
12. Corner Gammon
13. Slipper
14. Middle Gammon
15. Gammon Flock

Ham and Gammon: Ham and gammon come from the hind leg of a pig Gammon is cut off the side of a cured porker.

Ham, on the other hand is cut off at the fresh pork stage and includes a substantial part of the loin. It is cured after cutting, usually by the long process of dry salting.

Cuts and their Uses

	CUT	USE
(i)	End collar	Boiling; Braising
(ii)	Fore hock can be bought boned and rolled	Boiling
(iii)	Prime collar	Boiling (either whole or cut into smaller pieces)
(iv)	Top streaky used for frying and grilling. Can be boned and boiled.	Thinly sliced rashers can be
(v)	Top back (good lean cut) in thin rashers and grilled.	Boiled or braised whole or cut
(vi)	Back and ribs	Cut into regathers and grilled.
(vii)	Prime streaky (the best cut of streaky bacon).	Boiled in one piece or cut into rashers.
(viii)	Short back (one of the best cuts for breakfast frying)	Frying

Contd...

	CUT	USE
(ix)	Thin streaky Frying (Crisp and well cooked)	Grilling
(x)	Long back or grilled quickly.	Sliced thinly and fried gently
(xi)	Flank (good cut to be served with liver or other meats and in a minced bacon recipe)	Boiling Frying
(xii)	Corner gammon (lean and well flavoured) rashers for grilling.	Boiling whole (serve hot or cut into
(xiii)	Slipper (gammon lean small cut)	Boiled whole or cut into rashers.
(xiv)	Middle gammon or lean rashers for frying.	Boiled whole (serve hot or cold)
(xv)	Gammon hock baked (serve hot or cold.)	Should be partly boiled and then

Beef—Signs of Good Quality:

(i) The lean meat should be bright red with small flecks of white fat.

(ii) The fat should be firm, brittle in texture, creamy white in colour and odourless. Older animals and dairy breeds have fat which is usually deeper yellow in colour,

(iii) Beef in good condition should have streaks of fat running through the lean section and the flesh should rise again quickly after being pressed with the fingers. Any cut surface should present a slightly moist appearance to the touch.

Order of Dissection: A whole side is divided into the forequarter and hindquarter, the division being between the wing ribs and fore ribs.

Hindquarter:

(i) Remove the rump, suet and kidney.

(ii) Remove the thin flank.

(iii) Divide the loin and rump from the leg. (The leg will consist of topside, silverside, thick flank and shin)

(iv) Remove the fillet.

(v) Divide the rump from the sirloin.

(vi) Remove the wing ribs.

(vii) Remove the shin.

(viii) Bone out the aitchbone.

(ix) Divide the leg into topside, silverside and thick flank.

Forequarter:

(i) Remove the shank.

(ii) Divide in half down the centres.

(iii) Take off the fore ribs.

(iv) Divide into joints.

Joints and their Uses

	JOINTS French	USES	APPROX WEIGHT
(i) Shin	Jambe de derriere (Jambe de devant)	Consomme beef tea, stewing	4.00 kg.
(ii) Topside	Culotte de boeuf	Braising, stewing, second-class roasting	7.5 kg.
(iii) Silverside	Gite a la noix, semelle	Salt beef	10.00 kg.
(iv) Thick flank	Gite a la noix	Braising and stewing	6.7 kg.
(v) Rump	Rump steak, colutte de	Grilling and frying as steaks or braised boeuf	7.2 kg.
(vi) Sirloin	Aloyau	Roasting, grilling and frying	10.00 kg.
(vii) Wing ribs	Cote d'aloyau	Roasting, grilling and frying steaks	5.9 kg.
(viii) Thin flank	Bavette	Stewing, boiling, sausages	4.00 kg.
(ix) Fillet	Filet	Roasting, grilling and frying	2.72 kg.
(x) Fore rib	Cote premiere	Roasting and braising	6.5 kg.
(xi) Middle rib.	Cote deactivate	Roasting and braising	8.00 kg.
(xii) Chuck rib	Cote du collier	Stewing and braising	4.08 kg.
(xiii) Sticking piece	Cou (collier)	Stewing and sausages	10.00 kg.
(xiv) Brisket	Poitrine	Salted and boiled	6.8 kg.
(xv) Plate	Plat de cot	Stewing and sausages	5.9 kg.
(xvi) Leg of	Gigot mutton	Braising and stewing	10.5 kg.
(xviii) Shank fat,	Jarret bone of kidney	Consomme, beef tea (trumeau)	4.08 kg. 7.2 kg.

N.B.: Weight of aitchbone x 10 = total weight of bones in hind quarter.

Minute Steak: Sirloin cut into 1.25 cm. (½")thick slices, flattened with a cutlet bat dipped in water to make the steak as thin as possible, trimmed and grilled or fried.

Sirloin Steaks: Sirloin cut into 1.25 cm. (½")thick slices, trimmed and fried or grilled.

Double Sirloin Steaks: Sirloin cut into 2.5 cm. (1") slices, trimmed and fried or grilled.

Porterhouse and T-bone steak: Complete slices of sirloin including the bone and fillet.

Chateaubriand: This is a double fillet steak taken from the head of a fillet of beef, 3.5 to 10 cm. (1½"-4) thick, with an average weight of 340 gm. to 1 kg. It is generally served grilled at a temperature of 204.5°C (about 400°F), done to the taste of the customer.

Fillet Steak: These are approximately 115 gm. to 170 gm. in weight cut into 2.5 cm. (1") thick slices trimmed and grilled or fried.

Tournedos: Taken from the middle portion of the fillet. Remove all fat.

Fillet Mignon: Taken from the tail end of the fillet. All fat and sinews are removed and it is minced or sliced as required.

Entrecote Steak: A slice of steak cut between two bones of a rib of beef, or cut from the contrefilet (bone sirloin or undercut), and weighing about 650 gm.; sufficient for four persons.

Entrecote Minute: Cut as above from the sirloin or contrefilet as thinly as possible, about 1 cm. (2/5") in thickness, well flattened, and served grilled or sauteed.

Benefits of Chicken

Choosing Poultry: Only young birds are suitable for roasting; older birds may be boiled , steamed, braised or stewed. A roasting chicken should be about nine months to a year old; geese and other birds for roasting should not also be older than one year.

Signs of a Young Bird

(i) The feathers, especially the quills on the wings, should be easy to pull out.

(ii) There should be down all over the body, but especially under the wings.

(iii) No long hair on the body.

(iv) The skin should be white or clear and smooth.

(v) The feet should be supple, with smooth, even, overlapping scales,

(vi) The comb and wattle should be small and not well developed.

Signs which Show that a Bird is Fresh:

(i) Feathers - light and fluffy,

(ii) Eyes - prominent and clear.

(iii) No marked or unpleasant smell,

(iv) Feet - moist and not stiff and dry

(v) Skin - clear with no dark or greenish tinge.

Points which Denote Good Condition in Ducks: The breast should be plump. The bill and feet are yellow in young birds, but they darken with age. The webbing of the feet should be smooth, tender and easily broken, and the underbill soft and pliable.

Preparation

Storing poultry: Fowls and ducks are best used fresh and should not be stored longer than 1 or 2 days before cooking. Protect them from flies and hang by the feet in a dry airy place.

Plucking: If poultry is plucked while still warm the feathers are easier to remove.

(i) When plucking, do not sit in a draught or the feathers will blow about.

(ii) Put the bird on a large sheet of strong paper.

(iii) Take care not to tear the skin, especially when plucking game birds.

(iv) Hold the bird firmly with the left hand and pluck with the right.

(v) Pluck out small handfuls at a time, starting from under one wing then continue plucking until one side is completely clear.

(vi) Remove all the pin feathers with a knife.

Singeing: A bird can be singed in either of the two following ways. Care should be taken not to scorch or blacken the skin.

(i) Hold the bird by its head and feet, and quickly and carefully singe off the hairs over a small gas jet.

(ii) Hold the bird in the left hand and singe with a lighted taper or paper spill.

The Giblets: These can be used for making stock for gravy to serve with the roast bird, for enriching the stock when the bird is boiled, or for making giblet pie or giblet stew. The liver and kidneys are not boiled with the fowl, as they tend to darken the meat.

They may be half-cooked, chopped and used as part of the stuffing or cooked in other ways. Instructions for cleaning giblets are given below.

Liver: Carefully cut away the gall bladder, taking great care not to break it.

Stomach: Remove any fat and reserve it. Scrape the gizzard lining well.

Heart: Cut free of tubes, cut open, and scrape away all congealed blood.

Kidney: With a large bird, such as a turkey or goose, these are large enough to be used. Cut them open and cut out the core.

Neck: Scrape away all congealed blood.

Wash the liver, gizzard, heart, kidney and neck in salt water and rinse well.

Drawing the Tendons: If a bird is not young and tender, it is advisable to draw the tendons from the legs before roasting.

(i) Carefully cut through the skin of the leg at the knee joint, taking care not to cut the tendons at the same time,

(ii) Break the joint and swing the leg free to expose the tendons,

(iii) Pull out the tendons one at a time with a strong metal skewer. There will be 7 of them in each leg.

Stuffing: Birds can be roasted, stuffed or unstuffed, but they are usually stuffed. Stuffing can be put inside the bird only, or both inside and at the crop, covered by the skin of the neck.

Veal forcemeat may be used for the inside and sausage meat for the crop or vice versa. Pack the veal forcemeat, or any similar stuffing, fairly loosely to allow for swelling.

Trussing:

Method (A)

A trussing needle and some thin, clean string will be necessary. A trussing needle is a strong steel needle about 23 cm. (9") long with a large eye.

(i) Fold the skin of the neck over the back of the bird, and cross the tips of the wings over, to keep it securely in position.

(ii) Turn the bird breast upwards and put the tail through the vent.

(iii) Thread the trussing needle and pass it just under the top joint of the right wing through the bird and out through the same joint in the left wing. Run the needle back through the bird so that it enters and comes out just above the bottom joint of the wings.

(iv) Tie the ends of the string securely in a bow, taking care not to drag the flesh.

(v) Re-thread the needle, press the legs well into the sides towards the wings, plumping out the breast, and run the needle through the back of the bird, just under the thigh joint, and return through a point near the ends of the legs. Tie both ends of the string firmly together.

Method (B)

(i) Fold the skin of the neck over the stuffing and secure firmly with a small skewer.

(ii) Turn the bird breast upwards and put the tail through the vent.

(iii) Pass a large skewer through the wing pinions to hold them securely in position,

(iv) Tuck the legs well into the sides of the bird to plump out the breast, then tie the legs firmly together with a string; pass the string underneath the bird, cross it, then bring it up again and twist it round the wing. Skewer and tie firmly.

Trussing a Fowl for Boiling:

(i) Cut the skin of the legs all round at the knee joint, crack the joints and pull the shanks off with as many of the sinews as possible.

(ii) Thread the trussing needle and secure the wings in the same way as for the roasting bird, or fix them securely with a skewer,

(iii) Carefully loosen the skin round the legs, and slip them into the body of the bird. Secure in position with the trussing needle and string or with a skewer.

Importance of Fresh Vegetables

Test for Freshness: All vegetables, particularly green ones, are best when taken straight from the garden. But this is not always possible so it is essential that one should know how to judge a well-grown vegetable. A well-grown vegetable is usually tender and free from pests.

Vegetables are classified into three divisions: (i) Root vegetables, (ii) Leafy vegetables, (iii) Other vegetables grown above the soil.

Root Vegetables: These should be firm and heavy for their size. They should be free from earth and grubs. Carrots should be heavy, smooth skinned and orange in colour.

Leafy Vegetables: These should be crisp and give a crunchy sound when squeezed. Break the vein of a leaf; it should give a sharp snap. It the leaves are limp, then the vegetable is not fresh.

Other Vegetables: Cabbage and cauliflower should be free from greenfly. The cabbage should have a firm heart and the outer leaves should stand the same test for freshness as green leafy vegetables. Cauliflower should have a close, white flower. Tomatoes should be of a rusty red colour and should not be soft. Runner beams and French beans should not be stringy. Peas should be full in the pod, and green. Ladies fingers should snap if the ends are broken. Brinjals should have a smooth skin and should not be heavy for their size.

Food Importance

Vegetables play an important role in the human diet. Properly chosen and eaten, cooked or raw, they make an invaluable contribution toward the supply of vitamins and minerals. Root vegetables are good sources of carbohydrates, and pulses (peas, beans, lentils, etc.) of vegetable proteins. A good balance of ingredients essential for growth and maintenance in human beings is more easily attained by a diet made up of both plant and animal life. Vegetables also provide variety in a meal, help to make a meal attractive by introducing colour, and furnish roughage.

Mushrooms have been found to contain one or more antibiotics, one of which is effective against tuberculosis and germs which cause boils. Mushrooms which have always been a food favourite are rich in vitamins and low in calories. Seven average sized mushrooms have only 1-2 calories.

Whenever possible at least two vegetables should be served in addition to potatoes. Salads should be made popular and should be served at both meals.

Salads are dishes made up of meat, poultry, game, fish, shellfish, eggs, vegetables, fruits and milk products—normally served cold but can be served hot or frozen. They can be made out of a single ingredient or a combination of ingredients. Salads are normally served as an accompaniment to the main course, a meal by itself or a sweet course (dessert). There should be special stress on hygiene as most of the ingredients are served raw. The ingredients used and the colour, flavour, texture and presentation may be varied according to the purpose.

There are four basic parts in a salad:

The Underliner: These are generally greens either shredded or in large pieces. Tear the greens instead of cutting with a knife. The greens must be crisp and chilled. To keep the greens crisp, it is necessary to store them in a manner that retains the maximum amount of water in the cells. This could be achieved by storing them at refrigeration temperature in the hydrator drawer wrapped in a moist cloth or in a covered container. The loss of water in cells can be reversed. To do this reduce the temperature to 2-3°C (about 34°F), and moisten the greens. When the cells are once again filled with water the greens will be crisp.

The Body or Heart of the Salad: This will form the major component and can be made of one ingredient or a combination of ingredients. Ingredients should look sparkling fresh with each individual piece intact. Fruits are particularly fragile and should be handled as little as possible. Dressings These make the salad more appetising although diet conscious people today eat their salads without a dressing.

Dressings: Dressing must be carefully selected. Depending upon the salad, the dressing could be either basic French or an emulsion. Pleasing flavours can be added to and variations made in the dressings by adding ingredients like cayenne pepper, tabasco sauce, diced hard boiled eggs, assorted herbs, a variety of cheeses, chutney, etc. Dressings should enhance the salad and not obscure it.

Garnish: The garnish should be the focal point and should be carefully selected for its simplicity and to enhance eye appeal and flavour. Edible garnish is more appropriate than inedible ones. Sometimes the dressing is also used as a garnish by piping it in a decorative form on the salad.

Special stress should be given to the quality of ingredients, eye appeal, simplicity, neatness and contrast or harmony in colour and in texture. All ingredients should be edible, clean, fresh and free from defects, such as bruises, rotten spots, insects, dirt, sand and insecticide. Raw fruits and vegetables should be at their optimum stage of maturity. All inedible Portions should be removed. Remove any excess moisture from ingredients as it thins the dressing, making it less effective.

The aesthetic qualities are determined by the form and shape of the food as well as by the colour. Skilfulness and artistic cutting and arranging of salad ingredients can create an appealing effect.

Try for an uncluttered appearance.

Always prepare salads as close to time of serving as possible.

Use of Vegetables

They should be washed before peeling. A knife or a peeler may be used.

The purpose of cutting is to ensure uniformity of size and even cooking.

The types of cutting used for vegetables are dicing, chopping, grating, slicing, shredding, cutting into rings, mincing, etc. The object of preparation is to remove spoilt, soiled and indigestible parts.

Cooking: Fruits and vegetables are cooked for many reasons:

(i) To preserve them,

(ii) To soften the product,

(iii) To improve the flavour.

(iv) To increase the digestibility.

Whatever the methods of cooking applied, the objective is to lose as little as possible of the natural flavours, colour and nutrients.

Cooking affects the palatability of the vegetables, *i.e.*, their texture, flavour and colour. While palatability may not affect the nutritive value of vegetables, it affects their desirability.

All the different methods of cooking can be effectively employed in the preparation of vegetables.

The effect of cooking on starch, protein and cellulose brings about the change in texture observed in vegetables when they are cooked. The predominant cause for change in texture will depend upon the composition of the vegetable. For example, with leafy vegetables which contain a high percentage of fibre, the change in texture is primarily due to the effect of cooking on the hemicellulose, whereas in potatoes the gelatinisation of starch is the chief factor. While this is true, it has also been noted that the change in any one ingredient is not the only cause that makes vegetables tender. For example, gelatinisation of the starch in a potato is completed before the potato is considered done.

As the amount of protein in vegetables is small, the effect of coagulation during cooking is not of great significance.

Alkalies such as baking soda aid in softening of vegetables as they speed up the solution of pectic substances and dissolve the hemicellulose. Acids, on the contrary, harden the vegetables, as acid precipitates pectin.

The texture of the cooked vegetables can be controlled to a great extent by the length of the cooking time. The longer the cooking period, the softer the product.

Flavour: The flavour of vegetables depends upon many substances—sugar, acids, tannings and a variety of volatile substances. There is a predominance of sweetness in beetroot, peas, corn and carrots, but each vegetable has other ingredients which contribute to its characteristic flavour and aroma. The pronounced bitterness or astringency of certain vegetables is due to the presence of tannin. The characteristic odour of the onion family—chives, shallots, onions, garlic and leek—is due to a readily volatile substance—allyl propyl disulfide. As this substance is soluble in water the longer the vegetables are cooked in water, the milder is the flavour.

The colour of the cooked product will depend on the kind and amount of colour pigment, the method of cooking and the acidity of the cooking medium. The colour pigments of fruits and vegetables are chlorophyll, carotenoids, flavones and anthocyanins. Tannings, though not colour pigments, undergo chemical changes during cooking and thus affect the colour of many cooked vegetables.

Chlorophyll: Chlorophyll is made up of two compounds, chlorophyll 'a' and chlorophyll 'b'. Both are organic esters and both contain magnesium. The reactions taking place during cooking affect these radicals. Chlorophyll is slightly soluble in water but dissolves easily in many organic solvents. Like other esters it is subject to hydrolysis, the reaction taking place in the presence of alkali, forming chlorophyll liens—salts of chlorophyll which are intensely green in colour.

In the presence of acids, chlorophyll undergoes another type of decomposition with the loss of magnesium from the molecule. The resulting compounds are called phaeophytins. They are yellow to brown in colour. Both reactions, with acids and with alkalis, are irreversible in the sense that chlorophyll cannot be formed again by treating the decomposed products with a neutralising agent.

Any cooking process which involves the use of acids or alkalies will affect the colour of green vegetables. During the life of the plant the chlorophyll is held in the cytoplasm of the plant cells in such a way that it is protected from the regions of changing acidity. When cell life is destroyed, the constituents of the cells intermingle and the green colour fades, becoming yellow and brown.

When a green vegetable is cooked the cell structure gets slowly destroyed and the acid of the cells gradually comes in contact with the chlorophyll. Hence green vegetables cooked for long periods lose their greenness. If the cooking pan is kept open during cooking, some of the acids of the vegetables may evaporate, preserving some of the green colour. When soda is added to the water in which green vegetables are cooked there is no loss in colour. The alkali neutralises the acidity of the vegetable and also speeds up the cooking process. Loss of colour also depends on the temperature of cooking. At high temperatures there is greater hydrolysis and, therefore, greater loss of colour.

Carotenoids: Carotenoids include two types of colour pigments—the hydrocarbons, *e.g.* carotene as in carrots, and the oxygenated hydrocarbons, *e.g.* lutein, the colouring matter in egg yolk.

These pigments like all other hydrocarbons do not dissolve readily in water. They are also not affected by acids or alkalies. The oxygenated carotenoids are also chemically resistant to acids and alkalies as well as insoluble in water. Other carotenoids are xanthophyll, associated with carotene, lycopene (the red pigment in tomatoes and watermelon), zeaxanthin (the yellow pigment in egg yolks and corn), capsanthin (in red pepper) and fucoxanthin (in brown seaweed).

Anthocyanins: Anthocyanins are colour pigments which are found more in fruits and flowers than in vegetables. The colour varies from red to blue. Red or purple cabbage, purple broccoli, beets and skins of aubergine and radish contain large amounts of anthocyanins. There are 28 known anthocyanins. They differ in their degree of solubility and sensitiveness to conditions which produce colour changes. They are inclined to be red in acidic and blue in basic solutions. The colour changes of anthocyanins, unlike those of chlorophyll, are reversible.

While cooking vegetables containing anthocyanins, the loss of colour as well as change of colour must be considered. Beets lose their colour or 'bleed' during the preparation. This is due to the great solubility of anthocyanins. High temperatures bleach this pigment, the colour returning partly when it is cooled.

Natural acids in fruits increase the redness of the cooked fruit. The colour of cooked red cabbage depends on the acidity of the cooking water. Tap water gives it a purple colour. Red cabbage is produced by adding acid to the water. The addition of cooking soda gives a bluish product.

Flavones occur in all plants, but particularly in white onions, white cabbage, cauliflower, etc. In an acid medium, they are colourless, while in a basic medium they are yellow. The reaction like that of arithocyanin, is reversible. Vegetables such as cauliflower can be made to change colour from white to yellow and back by holding them in fumes of acid or ammonia. Flavones are readily soluble in water as can be observed from the colour of the alkaline vegetable water, which is yellow. During cooking it is important to see that white vegetables do not turn yellow. The addition of acids such as lime juice or vinegar will keep them white.

Boiling: Vegetables are usually put into boiling liquid. This is to prevent the nourishment escaping into the liquid and also to curtail the action of plant enzymes which have harmful effects on vitamins. The exceptions to this are dried vegetables and root vegetables which should be started in cold water.

Salt is generally added to the water, except when cooking dry vegetables, which do not cook well if salt is added at the commencement. The lid is kept on generally, except when cooking green vegetables, where the volatile acids present in chlorophyll must be allowed to escape or else they will discolour the vegetables. Lime juice brings out the whiteness of cabbage and cauliflower if added before cooking.

Potatoes: Scrub and scrape new potatoes, peel and eye old ones. Keep under water to avoid discolouration. To boil, put in salted water and cook till tender, drain and use as required.

Roast potatoes: Parboil, drain and cook in dripping around the meat for at least 45 minutes.

Saute potatoes: Slice cold boiled potatoes and fry in very little fat till brown.

Potato chips: Two methods are used:

(i) Peel and cut them in to finger lengths. Parboil, drain and put into hot fat. Remove when half done. Put into smoking fat or boiling fat and brown just before serving.

(ii) Cut the peeled potatoes into finger lengths. Dry on a towel. Deep fry in a pan till a faint blue smoke rises. Fry chips until golden brown. Drain on absorbent paper, add salt.

Spinach: Remove the stalk and damaged leaves. Wash very thoroughly in several batches of water to remove grit and mud. Put into a pan. Add salt. Place over gentle heat, when the juice begins to run over spinach, raise temperature and boil for 15 minutes. Drain and chop or sieve as required. Add pepper and a small knob of butter to taste.

Before cooking spinach and other leafy vegetables, soak them for some time in water to which a little sugar has been added. This improves the flavour.

Green peas: Shell peas and place in boiling water with ½ teaspoon of sugar, salt and a sprig of mint. Boil till tender. Drain. Take out the mint and add a little butter.

French beans: String the beans. Slice them diagonally and cook in boiling salted water till tender (approx. 20 minutes). Cook with the lid off. Drain well and add a knob of butter.

Brussels sprouts: Found in a temperate climate. Cooked the same way as cabbage but kept whole. Slit stalk end into four before cooking.

Cabbage: Wash well, remove coarse leaves but use dark outer leaves. Cut in half. Remove the hard centre stalk, shred. Cook in a little boiling and salted water till tender (10 to 15 minutes approx.). Drain well. A few bacon rinds can be added to the water. Serve plain or tossed in melted butter.

Cauliflower: Wash thoroughly, remove outermost leaves if necessary but leave some on. Turn over and cut across the stalk. Put the cauliflower into boiling salted water and cook for 20 to 30 minutes.

Broccoli: Three types—white, green and purple.

Purple and green: To prepare, remove coarse stalks and wash the leaves and sprigs. Boil. Serve same way as cabbage or cauliflower. Particularly good with white saxice.

White: Prepare and cook as for cauliflower.

Turnips: Peel rather thickly. Cook in boiling salted water till tender; this takes 20 minutes to one hour. Drain, add white pepper and salt and butter. The cooked turnips can also be mashed up.

Vegetable Marrow: Very young vegetable marrow may be boiled whole and unpeeled. Cook in salted water for 15 minutes. Drain and serve with melted butter. Large marrows must be peeled and the seeds removed. Drain well and serve sprinkled with brown bread crumbs or with cheese sauce. Marrow may also be baked whole, (peeled or not) and stuffed with any savoury filling. Bake in a moderate oven at 177°C (350°F) for one hour. Cover the marrow with greased grease-proof paper.

Onion: Remove the root and the outer skin. Apart from being used as a flavouring agent, onions can be boiled and served as a vegetable or used stuffed.

Tomatoes: Grilled, fried, stewed, baked or raw.

Carrots: Scrub and scrape lightly. Peel old ones. Cut into strips or into roundels. Add just enough water, salt and a little butter. Cook till the carrots are tender and the water has evaporated. Serve sprinkled with chopped parsley. Young carrots can be boiled whole.

Corn on the Cob: Remove the sheath and silk threads. Cook in salted, boiling water till tender. Serve with melted butter.

Sweet Pepper: Capsicum, red or green. This can be served cooked or raw. To prepare the pepper, wash, cut out the stalks and remove the seeds. Red and green peppers may be cooked together to make a colourful vegetable dish. Slice and cook in a little salted water. Add a knob of butter. They may also be served stuffed, baked, boiled or stewed.

Celery: Wash celery thoroughly, using a brush to remove all soil. Separate the stalks. Tie into bundles. Cook in boiling salted water till tender. Drain and serve with white sauce.

Beetroot: Cut out the stalks about 2.5 cm. (1") from the root, taking care not to damage the skin. Wash and boil in salted water till tender (about 2 hours). Peel off the skin. Cut into cubes or slices and either serve hot, coated with white sauce, or cold in vinegar.

Leeks: Remove the coarse outer leaves. Cut out the roots and green top. Split the leaf end so that it can be washed thoroughly. Tie in bundles and boil in salted water. Drain. They are usually served with white or cheese sauce.

Broad Beans: Shell the beans. Cook them in boiling salted water and drain. Serve with parsley sauce. Skin older beans before serving.

Parsnips Peel thinly, using a potato peeler. Cut as required. Boil or bake.

Asparagus: Cut the woody portion, scrape the white part lightly. Remove any coarse spine. Tie in bundles. Place upright in a pan of salted water and boil for 10 minutes. Then lay the asparagus flat and cook till tender. Drain well, remove the string and serve with melted butter.

Artichokes: Only the lower parts of the leaves and lower sections are edible. Wash well. Cut off the stems and trim lower leaves if necessary. Place head downwards in boiling salted water and boil gently until tender, for about 30 minutes with the lid off. When cooked, the leaves should be easy to pull off. They may also be dipped in a thin batter and fried.

Serve hot with melted butter or margarine or hollandaise sauce. Do not pour the sauce over the artichokes. The lower sections without the leaves may also be used, either raw or cooked, in salads.

Artichokes (Jerusalem): Scrub well, peel, and to keep them white until ready to be cooked, place in a bowl of water to which a little vinegar has been added. Boil or bake. Serve plain boiled in the same way as potatoes or with white sauce. A very good au gratin can be made with these artichokes. Can be served in soups, separately or with another vegetable such as celery.

Colour Reaction of Vegetable Pigments

Pigment	*Vegetable*	*Colour in acid*	*Colour reaction in alkali*	*Colour reaction to metals*
Flavone	Cauliflower	Colourless (white)	Yellow	All-yellow Iron-brown
Anthocyanins	Red cabbage Beetroot	red	blue to green	Iron-blue Tin-purple
Carotenoids	Carrots	orange	orange	
Chlorophyll	French beans	olive green	bright green	Copper-bright
	Spinach			Iron-green

Pulse

Nutritive Function: Pulses are a good source of vegetable proteins and vitamin B complex. Being substances with very little taste, they should be combined with spicy and strong foods, *e.g.* bacon, ham, etc.

In Western preparations pulses are combined with onion, tomatoes, leeks, celery and bacon rind.

In Indian preparations pulses are combined with spices, coconut milk, tomatoes, leafy vegetables and herbs.

Cooking:

(i) If with skin, soak overnight in boiling water to soften skin, *e.g.* haricot beans, thick peas, etc.

(ii) Wash first before soaking so that the pulses can be cooked in the same liquid in which they are soaked. Those with no skins like lentils can be soaked in cold water. Bring to boil and cook with lid on.

(iii) Add salt after cooking, as salt inhibits the normal cooking process of lentils and pulses. Those with skins can be made into a puree, as it is the skin that causes indigestion.

In Western preparations these foods are rarely served plain boiled. They are generally mixed with other foods and browned by baking or frying. Use liquids for soups and gravies.

Cheese: Cheese, one of the oldest known food products, is universally popular. The chemistry involved in changing semifluid milk into semisolid cheese is complex and many factors enter into the final curing of cheese to produce the different varieties but they all begin with the action of rennet or lactic acid which performs the same basic function, that of coagulating the milk proteins.

Simply defined, cheese is the solid or casein portion of milk separated from the whey. The curdling of the casein is brought about by the action of rennet or lactic acid. The many different cheeses are the result of variation in the making and curing of each variety after initial curdling has taken place.

Many experiments and inventions in many countries of the world have produced cheeses ranging in texture from very soft to very hard and in flavour from extremely mild to very sharp. These cheeses are known by thousands of names, generally with names taken from the localities where they were first made.

Classification: Cheeses are classified into large general groups according to whether they are started by the rennet or lactic acid method. They may also be classified as unripened, mould ripened and bacteria ripened or as soft cheese, hard pressed cheese or blue veined cheese.

Cheese is one of the most highly concentrated of all protein foods. It is also readily digested. Experiments have shown that from 90 to 99% of all cheese is digested. It is also a complete protein. Since approximately 10 litres of fluid milk are required to make 1 kg. of cheese, cheese contains many of the nutrients of milk in highly concentrated form—milk proteins, fat, fat-soluble vitamins and minerals.

Wholemilk cheese contains the same properties as milk. Certain cheeses such as blue veined are made out of skimmed milk and, therefore, are less nutritious.

Normally cow's milk is used to make cheese, but certain well known cheeses are made from goat's milk. The texture of the cheese from goat's milk differs slightly from that of cow's milk. It is more crumbly.

Sheep's milk can also be used. The quality of the cheese depends to a great extent on the breed and the condition of the animal and the fodder given to it. Cheshire cheese is said to owe its fine flavour to the wild radish on which the cows feed, and its special nature is due to the mineral in the soil. Cheshire cheese, therefore, cannot be made in any other place as cheddar cheese can.

Certain cheeses develop a blue vein in maturing. This is sometimes a purely natural development, often sporadic and unpredictable. Sometimes fresh cheese is inoculated with pieces of blue cheese to catch the mould by contact. In some cases special bacilli are introduced. The mould is sometimes strengthened by brushing the cheese clean while the skin is soft, dipping it in whey and then rubbing it slightly with butter. This is done once a day for 10 to 20 days. Sometimes cheese is pierced with a copper wire. The main classes of cheese are:

(i) Unripened soft cheese, *e.g.* Cottage cheese, Cream cheese, Neufchatel.

(ii) Ripened in moulds by bacteria, *e.g.* Brie and Camembert.

(iii) Ripened by bacteria, *e.g.* Limbourger.

(iv) Semi-hard, ripened by bacteria in moulds, *e.g.* Gorgonzola, Roquefort and Stilton.

(v) Semi-hard, ripened by bacteria, *e.g.* Brick cheese, Munster.

(vi) Very hard cheese without gas holes, *e.g.* Cheddar, Edam, Gouda and Cheshire.

(vii) Very hard cheese with gas holes, *e.g.* Gruyere, Swiss Cheese, Parmesan.

Glossary:

Acid, acidity a description of a pleasant tang; it can be a defect if too pronounced.

Ammoniated a term describing cheeses that smell or taste of ammonia, a condition that affects the rinds of overripe cheese, primarily those with bloomy rinds such as Brie and Camembert. A hint of ammonia is not necessarily objectionable.

Annatto a yellow-orange dye extracted from the seeds of a South American plant and used to colour such cheeses as Cheddar, Momolette, Double Gloucester, Edam and many others.

Bloomy rind the white fleecy rind that develops on certain surface-ripened cheeses like Brie, Camembert, Double or Triple Creams and some Chevres. It is formed by spraying the surface of the cheese with spores of penicillium candidum while it is curing.

Chevres the French term for goat cheeses.

Gummy a negative term used to describe an overly plastic texture, as well as overripe rinds that have become sticky or gooey. Gumminess is undesirable in any context.

Salty most cheeses have some degree of saltiness; those lacking in salt are said to be dull or flat. Pronounced saltiness is characteristic of some cheeses, but over saltiness is a defect.

Springy a descriptive term for cheese with a resilient texture that springs back when gently pressed. Ripe or nearly ripe soft-ripened varieties should be springy.

Characteristics, Mode of Serving and Place of Origin of the Commonly Used Cheeses

	Cheese	*Characteristics*	*Mode Of Serving*	*Place Of Origin*
(i)	Cottage	Simplest of all cheese varieties. Uncured cheese made of cow's milk. Now made commercially from pasteurised skimmed milk to which lactic acid culture is added. Cream may be added if desired.	Salads, sandwiches, appetisers, cheesecakes and as paneer in Indian curries.	Uncertain,
(ii)	Cream Cheese	A soft uncured cheese made of cow's milk with cream added. The special richness an d smoothness comes from the whole milk with added cream. One of the few world cheeses made by the lactic acid rather than the rennet method of coagulating the curd. It is extremely mild in flavour.	Salads, sandwiches and appetisers.	Uncertain.
(iii)	Neufchatel	A soft rennet cheese made of cow's milk. Prepared in the same way as Cream Cheese with as lightly higher moisture content and lower fat content than Cream Cheese. Sometimes flavoured with pimento and other condiments.	Same as above.	France.
(iv)	Brie	Soft ripened cheese in small discs, made from cow's milk. One of the most famous and a great favourite in Europe.	Served after a meal and as a buffet cheese.	France.

Contd...

	Cheese	*Characteristics*	*Mode of Serving*	*Place of Origin*
(v)	Camembert	Soft ripened rennet cheese, made from cow's milk. Strong flavour like that of ammonia. It is a world favourite.	Served at the end of a meal with crackers, fruit, etc.	France.
(vi)	Limburger	Soft, highly flavoured cheese made from cow smilk. One of the most delicate and difficult cheeses to make.	Served at the end of a meal.	Luttich, Belgium.
(vii)	Gorgonzola	Hard blue-veined cheese made of cow's milk.	Served after meals, also used fors alads and b uffets.	Gorgonzloa, near Milan, Italy, France.
(viii)	Roquefort	Blue-veined, semi-hard. Has a sweet, piquant flavour. Only world-famous variety made from sheep's milk.	To be served at the end of a meal. Also used for salads and buffet serv-	England.
(ix)	Stilton	Semi-hard, blue-veined cheese. Made of cow's milk (whole with added cream)—spicy flavour. Surface rind is crinkled and brown with the interior creamy white and marbled by blue veins.	ices. Served at the end of a meal and with salads.	
(x)	Brick	Distinctly American cheese developed in Sand wiches and buffets. the mid-nineteenth century. Texture ranges from very firm to fairly soft and elastic. Flavours vary from extremely mild to sharp, depending on degree of curing.	Sandwiches and buffets.	America.

Contd...

Cheese	*Characteristics*	*Mode of Serving*	*Place of Origin*
(xi) Munster	Semi-hard, whole milk cheese. European Munster is highly flavoured and sharp. American Munster is mild in flavour, the curing time being shorter.	Sandwiches and buffets.	Munster (originally in Germany, now in france.)
(xii) Cheddar	One of the most popular and well-known cheeses of the world; made from cow's milk (whole or partly defatted). Hard and smooth. The English Cheddar is one of the oldest and best known of English cheeses. It is white in colour, close in texture and clean and mellow in flavour. American Cheddar, known as American Cheese, varies in colour from white to yellow. It is factory-produced, using lactic acid as a starter, with rennet added later.	Served after meals. Used in sandwiches and in preparation of foods.	Cheddar, England.
(xiii) Edam	Round, cannonball-shaped, bright red cheese. Produced by dipping ripened Holland. cheese in red-coloured paraffin. It is hard and rubbery in texture. In flavour it is like Cheddar, but milder and sweeter. It is made out of partly defatted cow's milk.	Served after meals of buffets.	Northern Holland.

Contd...

Cheese	Characteristics	Mode of Serving	Place of Origin
(xiv) Gouda	Semi-hard. It has the same flavour as Edam and is also dipped in red paraffin after ripening. It varies in size from Edam, weighing between 5-25 kg. (10-45 lbs.). Baby Goudas weighing about 500 gm. (1 1b) are now available. Goudas are made from partly defatted cow's milk.	Served after meals.	Southern Holland.
(xv) Cheshire	One of the oldest of English cheeses made of cow's milk. A hard cheese characterised by its loose, flaky crumbly texture ad solid in rich yellow colour or in its natural white colour.	Serve after meals, or in such dishes as Welsh rare bit. fondues, etc.	Cheshire County, England.
(xvi) Gruyere	Hard, with gas holes, and a nut-like, salty flavour. Made of cow's milk, which is usually partly defatted.	Served after meals.	Gruyere Switzerland
(xvii) Swiss cheese (known as switzerland.)	One of the world's most famous cheeses. It is hard, with gas holes, and a nutlike, sweet flavour. Made from partly defatted cow's milk (known) with gas holes, and a nutlike, sweet flavour. Made from partly defatted cow's milk. as Emmenthal in Switzerland	Served after meals, in sandwiches and with salads.	Emme Valley, italy.Emmenthal in

Contd...

Cheese	*Characteristics*	*Mode of Serving*	*Place of Origin*
(xviii) Parmesan	Hard, granular texture, sharp flavour. Made of partly defatted cow's milk.	Served after meals and for flavouring foods.	Parma and Lodi. Utakt.
(xix) Port-Salut	Semi-hard with rubbery texture and a flavour between Limbourger and Cheddar. It is made of whole slightly acid cow's milk. Oka cheese made in Canada belongs to the same family.	Served after meals	trappist Monasteries, France.
(i) Derby	Close buttery texture with a pale honey colour. It is mild when young but develops a fuller flavour as it matures.	served after meals and for flavouring foods.	England.
(ii) Lancashire	Crumbly texture. Has a mild flavour when young but develops a full and rather pungent flavour as it matures.	Ideal for crumbling over soups, hot-pos, or etc., Also well known for tosting quality	England.
(iii) Wensleydale	Flaky texture with subtle flavour. Clean, mild and slightly salty. It has an after taste of honey and is pale parchment in colour.	Used in preparation of foods.	Yrokshire, England.
(iv) Double Gloucester	Velvet texture like butter, resembles Cheddar in flavour and has a pale straw colour.	Served after meals or with fresh fruit salad and cream.	England.

Processed Cheese: Pasteurised processed cheese is the food prepared by mixing together, with the aid of heat and the addition of a small amount of emulsifying agent (not exceeding 3% of the total weight of the finished product), one or more cheeses of the same variety or two or more varieties. The emulsifying agents commonly used are sodium phosphate and sodium citrate.

Many cheeses are available in pasteurised processed form. Among the most common ones are American Cheddar, Swiss, Brick, Limburger, etc. Processed cheeses are widely served after meals. For cooking they have several advantages. They melt smoothly and quickly without separation or stringiness. They are also uniform in flavour and texture.

Cheese Cooking: Cheese cookery is an old art. Classic favourites like fondues, Welsh rare bit, macaroni and cheese, cheese pies and cheesecakes are well known. There are countless other dishes incorporating cheese which are of more recent origin. Cheese blends happily with many other food flavours and hence is a popular ingredient.

Cheese is a protein food and like all other proteins it is toughened by high heat. All cheese dishes should be cooked at low temperatures whatever the dish is. Whenever possible cheese should be melted in a double boiler or chafing dish rather than over direct heat.

Grate or chop cheese finely and dilute with some kind of starchy food such as flour, bread crumbs, macaroni, etc.

Add a small pinch of bicarbonate of soda ½ tsp to a kg. ½ tsp to a 1b). This softens the cheese, prevents stringiness and makes it more digestible.

Cook by moist heat whenever possible, or at least see that there is some moisture included in the dish. If and when possible, add the cheese only at the last moment after the rest of the food has been cooked. This is to prevent long cooking of cheese.

Up keep: The keeping quality of cheese varies greatly. Some of the hard varieties such as Swiss and Parmesan have a high keeping quality. Soft cheeses such as Camembert, Cream Cheese, etc. are highly perishable.

All cheeses require special care in handling once they are cut, for once cheese is cut it tends to dry rapidly. Mould may also form on any natural bulk cheese. This is in no way harmful, and can be cut off before being served without detriment to the quality of the remaining cheese.

Pasteurised processed cheese keeps well when leftover portions are rewrapped in the original paper and placed in the refrigerator. Bulk cheese or any hard cheese variety may be wrapped in wax paper, aluminium foil or plastic material. Tightly covered refrigerator dishes of plastic or glass are excellent for keeping cheese of all varieties.

All cheeses should normally be kept under refrigeration but before serving, they should be left out for a few hours. No cheese should be served chilled. The flavours of all cheeses are best when the cheese is neither cold nor over-warm.

Dishing and Garnishing: A good cook not only takes pride in preparing a dish correctly but also in presenting it properly to appeal to the different senses. This is done by suitable decoration and the way it is placed on the serving dish. One of the chief points that must be borne in mind when both dishing and garnishing is neatness.

Decoration that is done without restraint or badly carried out is worse than no decoration at all. Decoration done on hot dishes must be done quickly to prevent cooling of a dish which should be served hot. The colours of the garnish and the flavour should blend with the main dish. Proper garnishing has a three-fold role in food service. It helps "dress" the plate, "style" the product and adds colour, flavour and eye appeal of its own.

Accompaniments: Accompaniments are dishes such as sauces, salads or vegetables which are used to make a dish complete. They provide variety and improve the nutritive value of the meal.

What accompaniments should go with what, is usually determined by custom, based on a desire to enhance the flavour, to increase the digestibility and to improve a dish, *e.g.* roast beef with Yorkshire pudding, roast pork with apple sauce, roast duck with sage and onion stuffing and orange sauce, fried fish and chips with tartare sauce or tomato sauce, roast chicken with roast potatoes, buttered peas and bread sauce.

Forcemeat or stuffing are used to add flavour and piquancy to meat. They also help to increase bulk.

SAGE AND ONION STUFFING

Ingredients	*Quantity*
Onions	225 gm
Fresh bread crumbs	50 gm
Salt and Pepper	to taste
Butter	30 gm
Powdered sage or	
Fresh sage (chopped)	1 tsp
Method	

1. Boil chopped onions till tender, in just enough water. 2. Add butter and star till melted. 3. Add remaining ingredients, mix well and use as required.

FORCE MEAT STUFFING

Ingredients	*Quantity*
Fresh bread crumbs	115 gm
Suet, mutton dripping or Butter	50 gm
Grated lemon	½ lime
Chopped parsley	2 tsp
Mixed herbs	½ tsp
Salt and Pepper	to taste
Egg or milk to bind	

Method

1. Mix dry ingredients. 2. Add binding. 3. Mixture should be crumbly should not be pressed heavily and not mixed too wet.

BREAD SAUCE

Ingredients	*Quantity*
Milk	300 ml
Onion	10 gm
Margarine	30 gm
Fresh bread crumbs	50 gm
Cloves	2
Mace	a pinch
Lemon rind or a strip of lemon	
Seasoning	

1. Boil the milk. Infuse the hot milk with onion, cloves, mace and lemon rind for 30 minutes. 2. Strain over fresh bread crumbs, add seasoning and margarine; stir well to a smooth

consistency. 3. Stand in a warm place to thicken. Reheat and beat well; add one tablespoon cream if desired.

Sauces: Sauce is a liquid adjunct to a dish such as meat, poultry and vegetables used to moisten the food, to enhance the flavour, to provide contrast in taste or colour and sometimes to improve the digestibility (*e.g.* Roast pork with apple sauce); generally thickened by means of a liaison such as roux egg, blood, etc.

"A correct sauce is that wonderful production of the culinary art which forms a pleasant and exquisite accompaniment to all kinds of fish, meat, poultry, game and vegetables." The skill and knowledge of a cook is shown in no other part of culinary art so clearly and prominently as in the way the sauces are prepared. To make a perfect sauce is the height of the art of cooking.

The most simple dishes can be made appetising by the addition of a good plain sauce while the most recherche dishes can be improved and made still more palatable by a well-made sauce. Sauces in cookery are the essence of elegance. Every sauce, whether plain or rich, must possess a decidedly distinct flavour and character. Plain sauces should be simple and pure so that they taste of the materials employed, from which they take their name. Richer sauces always require a longer and slower process for their preparation.

Sauces were hardly made use of in English cookery till the beginning of the nineteenth century. Today there are at least 650 different types of sauces and gravies. A clever cook can devise as many varieties as he needs in sauces as an artist can paint pictures.

The art of sauce-making consists in preparing liquids from various materials by cleverly extracting and combining certain flavours into the liquid. Besides this, a good palate as well as the experience and skill of the most accomplished cooks and a thorough knowledge of the taste of those for whom the dish is being prepared, are necessary.

All sauces should be smooth, glossy in appearance, definite in taste, and light in texture.

Gravy is not a sauce. It is generally the juice of meat or vegetables seasoned but not thickened like a sauce. According to textbooks on cooking, a gravy may be called a sauce but a sauce cannot be called a gravy. Many of the more elaborate sauces have gravies as their foundation ingredient. It is easier to call a thinner liquid a gravy and a thicker one a sauce.

Liaison

The various processes of thickening sauces as well as soups are called "Liaison'. There are six distinct methods known for thickening sauces:

(i) Liaison with roux.

(ii) Liaison with egg.

(iii) Liaison with butter and cream.

(iv) Liaison with kneaded butter and flour (manied).

(v) Liaison with blood.

(vi) Liaison with farinaceous products.

Roux: The most popular and the most widely used thickening is roux. Roux is a mixture of flour and butter cooked to a definite degree - white, fawn or brown colour. Flour and butter are generally used in the same proportion. Roux can be prepared beforehand and stored, but it should be kept in airtight containers.

Approximately 60 gm of roux is required for 60 ml of liquid. If the roux is used cold, the stock used can be hot or cold. If mixed hot, the liquid, which should also be hot, must be gradually added to the roux away from the fire. This is then stirred over the fire till it boils. Roux should be allowed to cool slightly before the liquid is added to prevent the formation of lumps. All roux must be stirred constantly during the process of cooking.

White roux or Roux Blanc': This is a mixture of flour and butter cooked in a pan over a moderate fire without allowing it to get discoloured.

Blond roux or 'Roux Blonde': This is prepared by melting a certain quantity of butter and stirring in the same quantity or a little less of flour and by cooking it over a slow fire or in the oven till it gets a fawn colour.

Brown roux or 'Roux Brun': This is generally known as stock roux and can be prepared in bulk and stored. It is prepared more or less in the same way as 'Roux Blonde' but is cooked for a longer period to get a deeper brown colour. It is better to finish a brown roux in a slow oven so as to make it more brown without burning. This will also improve the aroma. With brown and blond roux, a mirepoix is added to give the necessary flavouring to the sauce. Mirepoix is an essence or extract of meat and vegetables. The following ingredients are used: 225 gm bacon, 1 carrot, 1 bay leaf, 1 sprig thyme, 2 small onions, and 2 cloves garlic. Chop all ingredients and fry without browning.

Egg Liaison: This is a thickening composed of egg yolks beaten up and diluted with a small quantity of cream, milk or cold white stock. Cream is used more often than stock. When the soup or sauce is boiling it is removed to the side of the fire and a spoonful of the boiling liquid is added to the egg mixture gradually and then the whole is poured slowly into the sauce and stirred over the fire without allowing it to boil. It is passed through a tummy before it is served. This liaison is used largely for banquettes, white ragouts (usually of vegetables) and white fricassees.

Butter and Cream Liaison: Butter and cream are incorporated in sauces and soups just before they are served. After the liaison has been added the sauce or soup is stirred vigorously without reheating. The flavour of the sauce alters if butter and cream are added too soon or if the sauce is allowed to boil after the addition of the liaison. The same applies to a plain butter liaison as well. Butter is added in small amounts the moment the sauce is taken off the fire. It is then stirred with a whisk and served without being reheated.

Kneaded Butter and Flour Liaison: Knead as much flour into the butter as it will absorb, to form a soft paste. Mix in small portions into a hot thin sauce. Stir constantly till the butter melts.

Blood Liaison: This is mostly used with hare or other game. It is made by preserving the blood of hare or game to which is added a little vinegar to prevent clotting. It is then strained through a fine sieve and generally added to the sauces.

Farinaceous Liaison: Arrowroot, corn-flour, potato flour, rice flour or other farinaceous products are commonly used for thickening sauces. Dilute with a little cold milk, cold stock or water. Pour through a strainer into boiling liquid. Stir continuously until it boils. Simmer gently for 10 minutes.

Utility of Spices

Down through the ages India has been known as the land of spices. Sweet or pungent, aromatic and seductive spices have played a special part in the life, legend and ritual of the country. With the march of time, although there have been several changes in the vise of spices, their qualities for enhancing food still reign supreme.

The primary function of spices in Indian food is to improve the flavour of the dish. Many spices such as cloves, coriander, cumin seeds, cinnamon, etc., because of their volatile oil contents, impart various flavours to the food. These appeal to the sense of smell and render the food more palatable.

The primary quality the common man looks for in a food product is its organoleptic quality (or the sense of taste), rather than its nutritive value. As a result even highly nutritious food is not accepted unless it is adequately spiced. It is only where pleasure to the eye and palate meet that food becomes fully acceptable. The best chefs of the world consider flavourings and seasonings absolute necessities for achieving this objective.

The success of cooking depends largely on their aid. However, spices must be used with skill and, above all, sparingly. All palates do not crave highly spiced foods, yet most people demand that food should be adequately spiced, for the flavour of insipid food can be improved very much by the use of some suitable spice. Even the people of the West, particularly the English and Americans who have a tradition of bland food, are now introducing a variety of spices into their dishes and appreciating the added piquancy of the food.

Western palates, however, are sensitive, and therefore can appreciate the subtle flavours more readily than we do, as unfortunately, owing to the continuous consumption of over spiced foods, we have lost this power. Whenever possible, only such spices should be used as bring out the natural flavours of the main ingredient, rather than imparting a new one. Intelligence, carefulness, thorough, sound judgement, a steady hand and a keenly perceptive palate are qualifications every cook must possess in order to prepare food that is appetising and pleasant to the taste.

Food Importance

Besides enhancing the flavour and aroma of food, spices have a physiological action beneficial to our system. They act as a stimulus to the digestive system and help digestion in many ways.

Every spice used in the making of a curry is a preservative. All have some antiseptic value and many are carminatives, *i.e.* they tend to reduce flatulence as in the case of omum (ajwain) water given to babies. The traditional combination of spices used for certain dishes must have been made with this factor in mind, *e.g.* the use of ginger or mint with peas. Peas are flatulence-forming and ginger or mint counteracts this effect. Asafoetida, commonly used with pulses (dais) of various types functions in a similar manner.

Some spices are used to give colour to food, to improve eye appeal, *e.g.* turmeric in yellow rice, red chillies in vindaloo, rattanjog in roganjosh, etc. Spices also serve as a thickening agent in the preparation of curries, *e.g.* onion and poppy seeds.

Various Benefits

The therapeutic value of spices is well known. Spices and herbs play a very important role in some of the simple but effective home remedies. Almost every spice has medicinal properties. Investigations are being conducted scientifically to elucidate the therapeutic properties of the various spices. Many have already been accepted and are being used today, *e.g.* aniseed (saunf) which has been used from prehistoric times as a flavouring for cough mixtures and as a herb tea to soothe the nervous system and induce sleep.

Scientists have recently probed the effect of cloves on the digestion. Studies of clove oil have shown that it stimulates the flow of gastric juices and is non- injurious to the lining of the stomach. Garlic has been accepted in both India and China as one of the treasured spices and medicinal agents. It has been indicated in the treatment of numerous diseases such as haemorrhoids, rheumatism, dermatitis, abdominal pains, coughs, loss of appetite, etc. Although garlic is claimed to be useful in the treatment of such a variety of diseases, it is not popular because of its strong smell and pungency. Research has now made it possible to administer the active principle of garlic in a fairly concentrated form.

Ginger tea is used commonly to ease an ailing stomach. Nutmeg is used medicinally to cure biliousness, diarrhoea and headaches. The ancients prized the seeds of fenugreek because of its stimulating effect on the digestive system.

To enhance the flavour and aroma of foods, spices may be used whole or in pieces, as in pulaos and biryanis, or in powdered or ground form. Indian cooks are artists in the use of spices. Money may limit the variety of spices used by some, but others are limited only by their own creative imagination. While freshly ground masalas (mixed curry spices) are preferred to powdered masalas, owing to the shortage of domestic labour there is a definite trend today towards using spices mainly in powdered form.

Many combinations of spices to suit different tastes are marketed today as various forms of curry powders. When using spices in powdered form it is essential that the powders be extremely fine. Coarse powders do not blend well with the rest of the ingredients; this causes a separation thus allowing the spices to float. The fine powders should be mixed into a paste with vinegar or water and then fried well before meat or vegetables are added. Besides curry powders, practically all the dry spices such as coriander, turmeric, cumin, fenugreek, cloves, cinnamon, mustard, etc., can be cleaned, dried and powdered.

Kinds of Spices

As a result of modem techniques, moist spices such as onion, garlic and ginger are now obtainable in powdered form. This process, while retaining the original flavours, improves the keeping quality of spices, and makes transportation more economical and easy. Onion powder is used commercially in the manufacture of processed meats and sausages, and for flavouring vegetables and meat.

Garlic powder has many commercial uses and is also now being used by the housewife who wants a clean and convenient method of using this popular flavour. All powdered spices, however, must be kept in airtight containers to prevent spoilage and deterioration in aroma and flavour, caused by the evaporation of volatile oils. Ground masalas are also available in the market today. They are packed in plastic bags and sold, but the keeping quality of these is poor as they are susceptible to spoilage.

The liquid spices now available in the market as straight liquid spices, such as liquid red chillies, green chillies, ginger, coriander, cumin, garlic, nutmeg, cinnamon, cloves, cardamom, asafoetida, saffron, curry leaves, etc., or as ready-to-use compounds such as liquid masala for curries, pulao, mutton, fish, etc., have revolutionised the use of spices.

These liquid spices are extracts of indigenous natural spices and are the result of years of research. By the separation of the active principles from the residual matter, concentrations are prepared. These are easy to use and are being used widely in the commercial manufacture of sauces, fruit juices, ketchups, salad dressings, tinned pulaos, curries, etc.

They have a better keeping quality than curry powders, and mix readily with the rest of the ingredients, thus bringing a uniformity of colour and flavour to the food. They are clean, compact, suitable for transportation and storage and are available in glass bottles with droppers, ranging from 15 ml (½ oz) to 500 ml (1 lb). They impart original flavours and colours. Thickening agents such as ground onion, poppy seeds, coconut and coconut milk should still be added to maintain the desired texture of the food. For those spices where flavour is the primary function as in cardamom, cloves, cinnamon, nutmeg, etc., the liquid form can be used most effectively.

It also makes it possible to be able to use out-of-season flavours. For example, when ginger or green chillies are not available, as in the summer months, the same flavours can be readily obtained by using liquid spices. Provided they are used carefully, liquid spices do not in any way increase costs, while at the same time they save labovir and time.

Allspice: This is the berry of 'Eugenia Pimenta', a small tree grown in the West Indies. The berries are gathered when green and unripe and dried in the sun. Then they turn black. Large quantities of these are used for the manufacture of sauces and pickles sold commercially. The berries combine the flavour of cloves, cinnamon and nutmeg. This spice is also known as 'Jamaica Pepper'.

Cloves: A very old and important spice, the clove was known before the time of the early Egyptians. The word is taken from the Latin 'Clavus' and the French 'Clou' both meaning nail, which describes its shape. The clove tree (of the order of the myrtle) grows to a height of 9 metres. The unopened flower buds are carefully harvested. Upon browning, they lose half their weight.

Cloves are used in both Western and Indian dishes, both savoury and sweet, *e.g.* pulaos, cooked ham, sauces such as bechamel sauce and bread sauce, meat dishes, fruit salads, baked apple, etc. The oil of cloves is used in dentistry to soothe toothaches, and its antiseptic properties are well known. It is grown on the islands of Zanzibar, Pemba and the East Indies, West India and other tropical regions.

Nutmeg and Mace: These are the only known cases of two different spices from the same fruit. Nutmeg is grown in the Dutch East Indies and Grenade in the British West Indies. The tree begins bearing fruit at the age of 8 years and continues to yield for half a century or more. In its wild state the tree grows to a height of about 22 m. or 70 ft. The nutmeg fruit resembles an apricot in size and shape. What we call the nutmeg is the seed. It is protected by a thin shell; the shell has a coat of orangy flesh which dries into mace. Although the aroma of mace is similar to that of nutmeg, separate uses exist for each. Nutmeg is used in Western and Indian dishes in puddings, sweetmeats, for flavouring egg and milk and in some curries. Mace is also used to flavour sauces, stocks, sweet dishes such as halwas as well as in betel nut mixtures. Nutmeg is used grated fine and mace coarsely crushed. When buying nutmegs choose those which are round, compact, of oily appearance and heavy for their size.

Mustard: A product of great antiquity and grown in most parts of the world, mustard is used in both Western and Indian preparations. It is used in the West in powdered form, as a table condiment and for flavouring sauces, etc. The pungency is more fully brought out when mustard is moistened with water (prepared mustard). In India it is used whole for tempering dishes, and in pickles. It is also an important ingredient in certain masala powders, *e.g.* vindaloo.

Cinnamon: This is taken from the bark of an evergreen tree belonging to the 'laurel' family and is chiefly cultivated in Ceylon and the East Indies. First the outer bark is stripped of the three-year-old branches, and then the inner bark is loosened and dried. The best cinnamon should not be too dark in colour and should be as thin as paper.

It has a fragrant odour and its taste is pleasant and aromatic. It is used in both Western and Indian cookery, in cakes, buns, ketchup, pickles, in pulaos with cardamom and cloves, etc. It is used in the preparation of garam masala for curries.

Cardamom: This is the fruit of a reed-like plant, native of the mountains of the Malabar Coast, but cultivated in Jamaica as well. The fruit is a small pod and the seeds within the pods have a strong, sweetish flavour. The pods vary from 0.5-2 cm. 1/4-1") and more in length. The small-sized ones are considered the best. Rich curries like khorma include cardamom. Powdered cardamom is a big favourite in cakes, puddings, halwas, etc. Pulaos and biryanis also depend on this spice for their popular flavour. Tincture of cardamom is used in stomach medicines.

Pepper: This is obtained from the seeds or berries of the plant 'piper nigrum' which grows in Malabar and other parts of India. Black pepper is made from dried whole berries, and white pepper from the same berries with the husk dried and removed. Pepper is one of the most universally popular spices, and is used in every type of savoury dish to improve flavour and to add to its piquancy. It is also one of the table condiments.

Mignonette Pepper: This is ordinary white pepper coarsely crushed.

Long Pepper: It is a spice similar in taste to ordinary pepper but not so pungent.

Cayenne Pepper: Dried fruits of the capsicum plant grown in the Cayenne Islands. This is red in colour, resembling the red chillies of India without being as pungent. This is used in Western cookery where pungency is required, *e.g.* for devilled prawns.

Krona Pepper: Bright red pepper from the Hungarian paprika. It is much milder than cayenne pepper and forms a palatable seasoning for Western savoury dishes.

Red Chillies: Chillies are grown in large quantities in India and are used extensively in Indian savoury preparations. They also form an important ingredient of curry powders. Apart from being used as a condiment, they are also used in several medicinal preparations for the treatment of lumbago, neuralgia, rheumatism, etc. Although there are several varieties they fall under two main groups: Capsicum Anum and Capsicum Frutescens. The variety acuminatum (Capsicum Anum) is the principle source of commercially used red chillies.

Coriander: The fruit of a plant called 'Corianderum Sativum' of the parsley family. It is extensively used in the preparation of Indian savoury dishes and forms an essential ingredient of curry powders. It is used in the preparation of frankfurters (sausages) and by confectioners and distillers. The leaves are used for flavouring curries, in salads and chutney and as a garnish.

Turmeric: This is the aromatic root or rhizome of a plant grown in India and the West Indies. It is sold in root form and as powder. The hard resinous flesh of the dried root varies from a dark orange to deep reddish brown, but in the powdered form it exhibits a characteristic yellow colour. Turmeric is a member of the ginger family but has a different flavour. It is used in curry powder not only for imparting colour and flavour to curries but also because of its preservative qualities.

Game and meats that are dried are soaked in liberal quantities of turmeric and salt. Turmeric is used freely in Indian medicines too. Plasters of it are applied in cases of bruises; as a dry dressing on open wounds; paste of turmeric and neem leaves is used to heal smallpox marks, and so on. Turmeric has sometimes been used to replace the more expensive saffron as a colouring agent.

Ginger: This is the tuberous root of a plant 61 cm. (2ft.) or more in height and is one of the few spices that grow underground. Ginger is mainly produced in Jamaica and to a less extent in Sierra Leone, China, Japan and India. Because of its stimulating and digestive properties it is employed medicinally for dyspepsia and colic. It is also frequently used to disguise nauseous tastes in medicines. The most common use of ginger is in making masalas, soft drinks, pickles and preserves, and in confectionery such as ginger biscuits, gingerbread, snaps, etc. Crystallised ginger is made from the young roots and is a delicacy. The best grades selected are called 'stem ginger'.

You can make instant pickle by chopping fresh ginger and green chillies and adding lemon juice together with salt. It is delicious specially one day after preparation.

Cumin seed: This is a herb of the caraway type, producing seeds of a pleasant smell and flavour which are very popular. It is most widely grown in several European countries, India and Mexico. The cumin plant is less than 30 cm high. The seeds come from the fruit and their slightly bitter taste is said to stimulate the appetite and digestive system. Cumin is added to curry powders and used whole in savouries, biscuits and fried rice. It is also used in the manufacture of liquors, pickles and sausages.

Fenugreek: These are the dried ripe fruits of an annual plant grown in many European countries and India. The fresh leaves and tender stalks are used widely as a vegetable. Because of their stimulating effect on the digestion they are administered medicinally even to horses and cattle. The seeds are used in making curry powders and fried whole for seasoning pickles and vegetables. When ground with water, fenugreek becomes a slimy paste and so is used to give batters smoothness.

Aniseed: The seed is small, oval in shape, somewhat similar to caraway and it grows in China, Mexico, Peru, Argentina, Spain, Malta, Syria, Lebanon, India, Pakistan, Hongkong and Egypt.

The seed is small, oval in shape, somewhat similar to caraway and is usually a greenish brown in colour. It is used in Western preparations and in Indian dishes. It is identified with the licorice flavour. It is used in making licorice confections, in baking, in the cordial anisette and other liquors. Its digestive properties are well known and it can be eaten plain after a meal.

Fennel Seed: The dried fruit of a perennial herb of the parsley family, fennel is grown -in Europe, India, Lebanon and Argentina. The seed has a licorice-like flavour resembling anise. Fennel is used both whole and ground. Bakers use it whole in Italy to stud breads and rolls. It is also used in sweet pickles, in the seasoning of soups, fish dishes, sauces, and in South Indian meat curries. It is also attributed with medicinal properties.

Caraway Seed: Fruit of a perennial plant of the parsley family. It has a pleasant, slightly sharp flavour with a sweet undertone. It is a native of Europe. Caraway seed is mostly used whole and is the principal flavouring in rye bread, and is widely used in rolls, biscuit and cakes. It is also used in cheeses, sauerkraut, pickles, soups, meats and stews. It is the chief ingredient of 'Kummel Cordial'.

Aromatic Herbs and Plants: The ones commonly used are parsley, bay leaves, thyme, marjoram, sage, tarragon, chervil, chives, onions, shallots, garlic etc. The bouquet garni which is the mainstay of French cookery consists of sprigs of chervil, chives, thyme, bay leaves, tarragon and parsley.

Parsley: It possesses a wonderful quality of masking the taste of a stronger flavour; so do not use too much of it when flavouring a dish. It plays an important part in cookery. It not only gives a finishing touch to stews and soups and sauces, but it is also used a great deal for garnishing purposes. Powdered parsley is an excellent herb which imparts a delicate flavour. To make powdered parsley, infuse some parsley leaves in boiling water for a few seconds. Drain and put into a hot oven to dry. Put through a sieve and use as required.

Tarragon and Chervil: The leaves of chervil possess a flavour which is appreciated by many people. Tarragon belongs to the same family as wormwood and the botanical name is 'Artemisia Dracunculus'. Of all the pot-herbs, tarragon and chervil give the strongest smell. They are used a great deal in French cookery for entrees, sauces, soups, salads, chad-froids, etc. Tarragon leaves are also used for flavouring vinegar.

Thyme: It belongs to the same family as mint. The leaves are used fresh or dry for stuffings, soups, etc.

It possesses a highly aromatic flavour and should, therefore, be used sparingly. Lemon thyme is a smaller variety and has a strong perfume like the rind of lemon. Leaves of ajwain (omum) can be used as a substitute for thyme.

Burnet: This is not used much in cookery now except for salads. When bruised it smells like cucumber.

Capsicum: There are several kinds of capsicum cultivated in the East and West Indies and in America. The capsicum fruit is both pungent and stimulating. In Mexico the pods are called chili and they are used to make hot-pickle and chili-vinegar. It is the powder of the dried seeds and pods that goes to make cayenne pepper (as a substitute use Kashmiri chilli powder). Capsicums are considered to be very wholesome.

Savory: There are two varieties of this herb; one is called the "summer variety". It is taken from a flowery herb. Both varieties are used extensively in Western cookery.

Marjoram: There are four kinds of marjoram. The sweet or knotted marjoram, originally a native of Portugal, is the one which is commonly used. It is used for soups, sauces, stews, etc., and used either fresh or dry.

Mint: This belongs to a family of plants called Labiatae. The spearmint or the 'Mentha Viridis' commonly cultivated in gardens has a better flavour and is the one used popularly in cooking. It has the property of removing flatulence.

Bay Leaves: The leaves of the common laurel are used in cookery to give a kernel-like flavour to stocks, mirepoix sauces, custards, puddings, blancmanges, and to the milk which is used for mixing cakes. They are generally dried and used.

Basil: A favourite herb in French cookery. It has a flavour similar to that of cloves. It is used for making such things as mock turtle soup and clear soups made out of shellfish. It is also used for flavouring vinegar.

Onions: The name onion is given to all plants of the onion tribe, *e.g.*, leeks, garlic and shallots. The onion is, next to salt, the most valuable of all flavouring substances used in cookery. When onions, shallots or garlic are used in cookery they should be well blended with other flavours. Whenever onion is used as a condiment or seasoning and the article properly treated, the strong smell is removed.

Garlic:This consists of groups of several bulbs called cloves, all enclosed in one membranous skin. When used carefully and sparingly, garlic is an excellent condiment. It is considered to be very wholesome and is a stimulant.

Shallots: This bulbous root resembles garlic. It is a native of Palestine and was introduced to England by the Crusaders.

Carrots and turnips: Next to onions, these are considered to be the most important flavouring vegetables for soups and sauces. Besides being used for flavouring, carrots and turnips are largely used for garnishing certain dishes such, as ragouts, boiled meat, etc. They are also used as a vegetable and as a puree for soup.

It plays an important part in sauces. The rind, juice and the essential oils contain valuable properties. The rind or peel is used for flavouring a variety of dishes. Generally the rind is grated, but the best way is to peel the rind with a very sharp knife as thinly as possible without touching the white part. Lemon rind is preserved in sugar.

When sour limes are in season buy them in large quantities. Extract the juice and mix one cup of salt to every four cups of juice. Shake well and keep in an airtight bottle—under refrigeration it keeps for months. Use whenever fresh lime is required.

Vanilla: This was first discovered by the Spaniards. It is the fruit of an orchid. The best variety is grown in Mexico. It has a delicious flavour and is largely used for flavouring puddings, custards, liquor chocolate, etc. For flavouring purposes it is better to use the vanilla pod or vanilla sugar rather than the essence of vanilla, since the odour of the essence escapes rapidly.

Curry Powders: For some basic curry powders, see the recipes at the beginning of the Indian Cookery Section of this book.

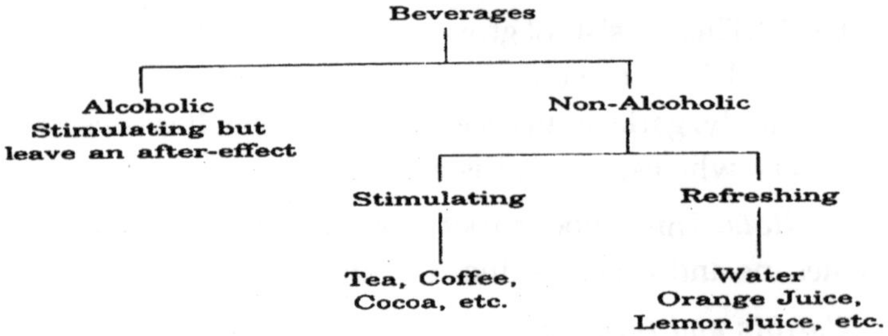

Various Tasks

To Prevent Water Depletion: Water being one of the essential constituents of beverages it is very useful in making good the loss of water content in the blood stream, specially after heavy physical exertion. Adequate supply of water is essential for the healthy functioning of all the tissues in the body. For an adult doing no manual work and when the weather is pleasant enough not to cause any excessive perspiration, three litres of water daily are necessary for good health. Two litres are obtained from food so one litre must be taken in the form of water and other beverages.

Refreshing and Restorative: Cold drinks in summer and hot drinks in winter have a greatly refreshing and restorative effect on the body particularly when fatigued.

Supplies vitamins: Certain beverages like lime juice, Ovaltine, etc., are a source of vitamin supply.

Stimulating: Alcoholic beverages are stimulating and have a definite value in case of illness, but if taken regularly even in moderate quantities, lessen the power of resistance to disease, and if taken in excess are very harmful. Beverages such as tea, coffee and cocoa besides being refreshing are also stimulating but they do not have any aftereffects.

Bibliography

Achaya, K.T. : *Indian Dairy Products,* Asia Publishing House, Bangalore, 1974.

Adam, H. : *Food Science,* John Wiley and Sons, New York, 1982.

Austin, A. and Ram, A. : *Studies on Chapati-Making Quality of Wheat,* Indian Council of Agricultural Research, New Delhi, 1971.

Bachman, W. : *Food Chemistry,* Maclaren and Sons Ltd., London, 1955.

Balsdon, J.P. : *Life and Leisure in Ancient Rome,* Bodley Head, London, 1966.

Beersteeker, E. : *The Biochemistry of Vitamins,* Reinhold, New York, 1950.

Bharathi, V.V. : *Anxiety Behaviour in Children,* Discovery Publishing House, New Delhi, 1992.

Board, R.G. : *A Modern Introduction to Food Microbiology,* Blackwell Scientific Publications, London, 1983.

Burton, T.L. : *Recreation Research and Planning,* Allen and Unwin, London, 1970.

Cambell, A.M., Penfield, M.P. and R.M. Griswold : *Experimental Study of Foods,* Houghton, London, 1990.

Cameron, Spencer : *Food Science,* Pargamon Press, New York, 1984.

Challoway, D.H. : *Nutrition and Physical Fitness,* W.B. Saunders Co., Philadelphia, 1979.

Cherry, G.E. : *Town Planning and its Social Context*, MacGraw-Hill, London, 1970.

Cherry, J.P. : *Protein Functionality in Foods*, American Chemical Society, Washington D.C., 1981.

Chichester, C.O. : *The Chemistry of Plant Pigments*, Academic Press, New York, 1972.

Dale, E. : *Management Theory and Practice*, McGraw-Hill, New York, 1973.

Dharamraj, E. : *Food and Hotel Legislations and Policies*, New Age Publishers, New Delhi, 1998.

Dumazedier, J. : *Towards a Society of Leisure*, Free Press, New York, 1967.

Dyke, S .F. : *The Chemistry of Vitamins*, Interscience Publishers, New York, 1965.

Edmunds, F. : *Environmental Administration*, McGraw-Hill, New York, 1973.

Eley, G. : *Wild Fruits and Nuts*, E.P. Publishing Ltd., Yorkshire, 1976.

Feiffer, M. : *Giving Places*, MacMillan, London, 1985.

Furia, T.E. : *Hand Book of Food Additives*, C.R.C. Press, Ohio, 1977.

Geiss, I. : *Proteins : Structure, Function and Evolution*, Menlo Park, California, 1983.

Gell, D. L. : *International Hotel Industry Policy*, Van Nostrand Reinhold, New York, 1990.

Gyorgy, P. and Pearson, W.N. : *The Vitamin*, Academic Press, New York, 1967.

Haschemeyer, A.H. : *Proteins : A Guide to Study of Physical and Chemical Methods*, John Wiley & Sons, New York, 1973.

Hudson, F. : *Development in Food Proteins*, Applied Science Publisher, London, 1982.

Inglett, G. : *The Quality of Foods and Beverages : Chemistry and Technology*, Academic Press, New York, 1981.

Jackson, R. : *The Geography of Recreation and Leisure,* Hutchinson, London, 1972.

Jasscler, D.K. and M.A. Joslyn : *Fruit and Vegetables Juice,* The AVI Publishing Co., Connecticut, 1971.

Johnson, A.H. : *Encyclopedia of Food Science,* The AVI Publishing, Connecticut, 1978.

Kacharoo, P. : *Pulse Crops of India,* ICAR, New Delhi, 1970.

Kadans, J.M. : *Encyclopedia of Fruits, Vegetables, Nuts and Seeds,* Parker Publishing Co., New York, 1973.

Kalyani, T. : *Home and School Environment,* Discovery Publishing House, New Delhi, 2003.

Kinton, R. : *Practical Cookery,* Edward Arnold, London, 1981.

Krause, M.V. and L.K. Mahann : *Food Nutrition and Diet Therapy,* W.B. Saunders Co., Philadelphia, 1979.

Kumari, P. : *Role of Anganwadi Workers in Integrated Child Development Services Programme,* Discovery Publishing House, New Delhi, 1994.

Lal, H. : *Total Quality Management : A Practical Approach,* New Age Publishers, New Delhi, 2000.

Law, C. : *Urban Tourism : Attracting Visitors to Large Cities,* Mansell, London, 1993.

Less, R. and E.B. Jackson : *Sugar Confectionery and Chocolate Manufacture,* Leonard Hill Books, Burks, 1973.

Lickorish, L. J. : *Hotel Industry and International Balance of Payments,* International Institute of Scientific Travel Research, Geneva, 1953.

Lloydryall, A. and W.T. Pendzer : *Handling Transportation and Storage of Fruits and Vegetables : Fruits and Peanuts,* The AVI Publishing Co., Connecticut, 1974.

Mackinney, G. and A.C. Little : *Colour of Foods,* The AVI Publishing Company, Connecticut, 1972.

Malhotra, R. K. : *Food Service and Catering Management,* Anmol Publications, New Delhi, 1998.

Marton, I.D. and A.J. Macbod : *Food Flavours*, Elesvier Scientific Publishing Co., New York, 1982.

Masefield, G.B. : *The Oxford Book of Food Plants*, Oxford University Press, London, 1975.

Mate, S.A. : *Bakery Technology and Engineering*, The AVI Publishing Co., Connecticut, 1972.

————— : *Cereal as Foods and Feeds*, The AVI Publishing Co., Connecticut, 1959.

Mayo, J. : *The Psychology of Leisure Travel*, CBI Publishing Company, Boston, 1981.

Mcwilliams, M. : *Food Fundamentals*, John Wiley & Sons, New York, 1979.

Medlik, S. : *Hotel Management: Past, Present and Future*, Heinemann, London, 1976.

Mellor, J. D. : *Fundamentals of Freete-Drying*, Academic Press, London, 1978.

Meyer, L.H. : *Food Chemistry*, Reinhold Publishing Corporation, New York, 1961.

Michael Eskin, N.A. : *Plant Pigments, Flavours and Textures*, Academic Press, New York, 1979.

Murthy, R.B. : *Breeding Procedures in Pearl Millet*, ICAR, New Delhi, 1977.

Nath, P. : *Vegetables for the Tropical Region*, Indian Council of Agricultural Research, New Delhi, 1976.

Oberleas, D. : *Trace Elements in Health and Disease*, Academic Press, New York, 1976.

Palmer, T. : *Understanding Enzymes*, John Wiley & Sons, New York, 1981.

Pamela, V.P. : *Directory of Wines and Spirits*, North Wood Publication, London, 1974.

Paul, P.C. and Palmer, H.H. : *Food Theory and Application*, John Wiley & Sons, New York, 1972.

Peckham, G.G. : *Foundations of Food Preparation*, MacMillan Publishing Co., New York, 1979.

Pigman, W.W. and Horton, D. : *The Carbohydrates : Chemistry and Biochemistry Plants*, C.S.I.R, New Delhi, 1969.

Pigram, J. : *Outdoor Recreation and Resource Management*, Croom Helm, London, 1993.

Potter, N.M. : *Food Science*, The AVI Publishing Co., Connecticut, 1978.

Pudney, John : *The Thomas Cook Story*, Michael Joseph, London, 1953.

Puruthi, J.S. : *Spices and Condiments*, National Book Trust of India, New Delhi, 1976.

Ranganna, S. : *Manual of Analysis of Fruit and Vegetable Products*, Tata McGraw-Hill Publishing Co. Ltd., New Delhi, 1984.

Rani, Usha D. : *Nutrition and Religion*, Discovery Publishing House, New Delhi, 2003.

Rao, D. V. : *Community and School Nutrition Education*, Discovery Publishing House, New Delhi, 1998.

Roberts, D.V. : *Enzyme Kinetics*, Cambridge University Press, 1977.

Rogers, C.A. : *Profitable Poultry Keeping in India and the East*, Treasure House of Books, Bombay, 1966.

Ross, G. F. : *The Psychology of Hotel Management*, Hospitality Press, Melbourne, 1994.

Schapira, J. : *The Book of Coffee and Tea*, St. Martin's Press, New York, 1975.

Schultz, G.E. and R.H. Schirmer : *Principles of Protein Structure*, Springer-Verlage, New York, 1979.

Shreedevi, V. : *Nutrition Education*, Discovery Publishing House, New Delhi, 1997.

Smartt, J. : *Tropical Pulses*, Longman Group Ltd., London, 1976.

Stadlmare, W.J. and. Cotterill, O.Y. : *Egg Science and Technology*, The AVI Publishing Co., Connecticut, 1973.

Stumbo, C.R. : *Thermobacteriology in Food Processing*, Academic Press,, New York, 2000.

Swern, D. : *Bailey's Industrial Oil and Fat Products,* John Wiley & Sons, New York, 1964.

Tandon, G.L. : *Preservation of Fruits and Vegetables,* Indian Council of Agricultural Research, New Delhi, 1960.

Tandon, H.P. : *Egg and its Care,* Ministry of Agriculture, Govt. of India, New Delhi, 1973.

Thangam, P.E. : *Modern Cookery for Teaching and Trade,* Orient Longmans Ltd., Bombay, 1983.

Thankamma, J. : *Food Adulteration,* S.G. Wasani, New Delhi, 1976.

Tressler, D.K. : *Food Technology the World Over,* The AVI Publishing Co., Connecticut, 1963.

Uma, K. : *Poverty, House Hold Fold Security and Nutrition in Rural Areas,* Discovery Publishing House, New Delhi, 2000.

Underwood, E.J : *Trace Elements in Human and Animal Nutrition,* Academic Press, New York, 1977.

Veldhnis, M.K. : *Food Science and Technology,* The AVI Publishing Co., Connecticut, 1977.

Vidya, D.B. : *A Textbook of Nutrition,* Discovery Publishing House, New Delhi, 1996.

Weiss, T. J. : *Food Oils and their Uses,* The AVI Publishing, Connecticut, 1970.

Whitaker, J.R. : *Principles of Enzymology for Food Science,* Marcel Dekker, New York, 1977.

Williams, P.N. : *The Chemical Constitution of Natural Fats,* Chapman and Hall, London, 1964.

Winitz, M. : *Chemistry of Amino Acids,* John Wiley & Sons, New York, 1961.

Yasoda, P. : *Determinants of Nutritional Status of Pre-School Children,* Discovery Publishing House, New Delhi, 1998.

Yukic, T.S. : *Fundamentals of Recreation,* Harper Row, London, 1963.

Index

D

E

F

G

❑❑❑